T0261345

Kuhn

Kuhn

Philosopher of Scientific Revolution

Wes Sharrock and Rupert Read

polity

First published in 2002 by Polity Press in association with Blackwell Publishers Ltd, a Blackwell Publishing Company

Editorial office:
Polity Press
65 Bridge Street
Cambridge CB2 1UR, UK

Marketing and production:
Blackwell Publishers Ltd
108 Cowley Road
Oxford OX4 1JF, UK

Published in the USA by
Blackwell Publishers Inc.
350 Main Street
Malden, MA 02148, USA

ISBN 0-7456-1928-2
ISBN 0-7456-1929-0 (pbk)

A catalogue record for this book is available from the British Library and has been applied for from the Library of Congress.

Typeset in 10½ on 12 pt Palatino
by SNP Best-set Typesetter Ltd., Hong Kong
Printed in Great Britain by TJ International, Padstow, Cornwall

This book is printed on acid-free paper.

Key Contemporary Thinkers

Published

Wes Sharrock and Rupert Read, *Kuhn: Philosopher of Scientific Revolution*

David Silverman, *Harvey Sacks: Social Science and Conversation Analysis*

Dennis Smith, *Zygmunt Bauman: Prophet of Postmodernity*

Nicholas H. Smith, *Charles Taylor: Meaning, Morals and Modernity*

Geoffrey Stokes, *Popper: Philosophy, Politics and Scientific Method*

Georgia Warnke, *Gadamer: Hermeneutics, Tradition and Reason*

James Williams, *Lyotard: Towards a Postmodern Philosophy*

Jonathan Wolff, *Robert Nozick: Property, Justice and the Minimal State*

Forthcoming

Maria Baghramian, *Hilary Putnam*
Sara Beardsworth, *Kristeva*
James Carey, *Innis and McLuhan*
George Crowder, *Isaiah Berlin: Liberty, Pluralism and Liberalism*
Thomas D'Andrea, *Alasdair MacIntyre*
Eric Dunning, *Norbert Elias*
Jocelyn Dunphy, *Paul Ricoeur*
Matthew Elton, *Daniel Dennett*
Chris Fleming, *René Girard: Violence and Mimesis*
Nigel Gibson, *Frantz Fanon*
Sarah Kay, *Žižek: A Critical Introduction*
Paul Kelly, *Ronald Dworkin*
Carl Levy, *Antonio Gramsci*
Moya Lloyd, *Judith Butler*
Dermot Moran, *Edmund Husserl*
James O'Shea, *Wilfrid Sellars*
Kari Palonen, *Quentin Skinner*
Steve Redhead, *Paul Virilio: Theorist for an Accelerated Culture*
Chris Rojek, *Stuart Hall*
Nicholas Walker, *Heidegger*

Contents

Acknowledgements

Thanks are due to Ralph Berry, Alexander Bird, Wil Coleman, Jeff Coulter, Alice Crary, Fred D'Agostino, Arthur Danto, Miguel De la Franca, Dave Francis, Steve Fuller, Ian Hacking, Richard Hamilton, Paul Hoyningen-Huene, Nick Huggett, David Johnston, Keekok Lee, Bob Lockie, John Preston, Lincoln Read, Angus Ross, Seamus Simpson, Richard Traub, Thomas Uebel and especially to James Conant, Stefano Gattei, John Haugeland, Vasso Kindi, Jehane Kuhn, Jay Elliot, Martin Gustafson, Clinton Tolley, and the many students to whom we have attempted to teach this material; to audiences at the British Philosophy of Science Society (Sheffield), at the University of East Anglia, the Indian Institute of Technology (Delhi), Binghamton University, Colgate University, Williams College, and the University of Paris; and to the understanding people at Polity Press.

Rupert thanks Glendra and Graham Read (for love and money), Emma Willmer (for love), and the University of East Anglia and the Arts and Humanities Research Board (for money). Wes and Rupert both owe much to John Lee.

This book has been our joint work throughout – together we have written, produced and agreed upon every line and the order in which our names appear on the cover is only thus because Wes, most of whose work is jointly authored, is fed up with nearly always being listed last due to his surname beginning with an 'S'.

We are grateful to the editors of the *Journal for the General Philosophy of Science*, of the *Harvard Review of Philosophy* and of *Wittgenstein*

Studies, where earlier versions of chunks of this book were published, and to Professor Ranjit Nair for comments. Acknowledgement is also due to the University of Chicago Press for permission to quote from Thomas Kuhn, *The Structure of Scientific Revolutions*, 3rd edition (1996).

Abbreviations for Kuhn's Works

BB   *Black-Body Theory and the Quantum Discontinuity, 1894–1912*
CR   *The Copernican Revolution*
ET   *The Essential Tension*
RSS   *The Road since 'Structure'*
SSR   *The Structure of Scientific Revolutions*

Introduction

The Legendary
Thomas Kuhn

Thomas Kuhn died in 1996, convinced that his lifework had been misunderstood, and failing to complete a categorical restatement of his position before his death. We think Kuhn was right, and will here try to set that situation right. If Kuhn was perhaps the most influential philosopher writing in English since 1950, even the most influential academic, the fact that he could rightly feel that both friends and foes misrepresented him implies a sad comment on the state of contemporary intellectual culture.

Kuhn, born in 1922, trained as a physicist, then turned to the history of science, and then to philosophy of science. He published four books during his lifetime. *The Essential Tension* (1977) was a collection of key papers. His first book, *The Copernican Revolution* (1957), and his last, *Black-Body Theory and the Quantum Discontinuity, 1894–1912* (1979), were historical studies of revolutionary periods in astronomy and atomic physics respectively. Immense success came with his long-delayed second book, *The Structure of Scientific Revolutions* (*SSR*) in 1962, a general philosophical interpretation of the long-term development of physical sciences in the West. The book attracted widespread attention, sold about a million copies, and became one of the most frequently cited contemporary philosophy texts. It remains, today, as controversial as on first publication.

Kuhn did not attribute his success to his own achievements. He was only one of a number of people thinking the same kinds of thoughts, including the French philosophers/historians of science Gaston Bachelard and Georges Canguilhem, and a small number of

Anglo-American philosophers including Mary Hesse, Stephen Toulmin and Norwood Russell Hanson, as well as their German precursor, Ludwig Fleck, whose *Genesis of a Scientific Fact*, published in 1932, Kuhn acknowledged had partially anticipated his own work. The time was right, Kuhn expressed these new ideas accessibly, and the zeitgeist took over.

To call Kuhn the most influential philosopher writing in English since 1950 is all the more striking since a substantial part of his work was in the history, rather than the philosophy, of science. If one excludes the illustrative examples from the key work, *The Structure of Scientific Revolutions*, then its argumentative content is scarcely ten thousand words. Further, though Kuhn's work was in either history or philosophy of science, his influence has been extraordinarily widespread, affecting the mainstream of philosophy as well as a host of disciplines in the social sciences and humanities, and beyond. Steven Weinberg, a leading physicist, rightly remarked that Kuhn's historical work alone would never have earned him the household name status he has among the chattering classes.[1] Whether that 'influence' has been for good or ill is much debated.

Kuhn's dissatisfaction with his own legend

Kuhn was disappointed that even those well-disposed towards him misconstrued his work, and he found it necessary to dissociate himself from many of his self-appointed followers, but this does not mean he was singularly hard to please. Kuhn was as critical of himself as of others, and over the years, made repeated attempts to restate and revise his work to his own satisfaction. A letter from Kuhn to the philosopher Guy Robinson expresses clearly Kuhn's sense of having found it hard to make out *himself* what he was trying to say: 'You've seen to an almost unprecedented extent what I've been up to. I couldn't have identified my position so clearly at the time I wrote *Structure*.'[2] The challenge of getting Kuhn right, then, is not a small one, and confronts us as much as the army of commentators and critics.

It is important to recognize that the connection between Kuhn's *history* and *philosophy* of science is much stronger than is usually appreciated, even when Kuhn is considered a founder of the 'historical approach' to the philosophy of science. If we seek the origins of Kuhn's arguments about the philosophy of science in his studies in the history of science, we get a much better grasp of his meaning.

Kuhn's reputation is the one that Steven Weinberg reiterates and decries:

> What does bother . . . on rereading *Structure* and some of Kuhn's later writings is his radically sceptical conclusions [*sic.*] about what is accomplished in the work of science. And it is just these conclusions that have made Kuhn a hero to the philosophers, historians, sociologists and cultural critics who question the objective character of scientific knowledge, who prefer to describe scientific theories as social constructions, not so different from democracy or baseball.[3]

Weinberg's remarks express animosities from 'the science wars', involving hostilities between the relatively new and burgeoning 'science studies' that recruit mainly from the humanities and social sciences, and those – philosophers and/or scientists – purportedly defending science against attack from science studies. Kuhn is seen as helping give science studies an initial impetus.

Weinberg is right in supposing that the reasons why he disapproves of Kuhn are the same ones that lead others to make a hero of him. Both 'friends' and 'foes' may endow Kuhn with the status of 'radical sceptic about science'. We cannot honestly say we find *nothing* in Kuhn that might indicate radical scepticism about science, but do say that it is to be found *only* when those ideas or remarks are taken in isolation. There is plenty of evidence that Kuhn was not sceptical at all (except about – the traditional – *philosophy* of science).

Weinberg complains that Kuhn makes change in science seem 'more like a religious conversion than an exercise of reason': when one scientific scheme displaces another 'it is not only our scientific theories which change but the very standards by which scientific theories are judged, so that the [scientific frameworks] which govern successive periods of normal science are incommensurable.' This means there is no common standard against which to match rival theories. There is, therefore, no sense in which theories developed through a scientific revolution can add cumulatively to what was known before. Replacement of one scientific theory by a newer one means that 'when new [scientific] problems arise they will be replaced [only] by new theories that are better at solving those problems' but this signifies 'no overall improvement'.[4] Kuhn *does* compare scientific change to conversion, but it does not follow that drastic conclusions are implied. Kuhn does not preclude the possibility of progress in science, but does deny that progress in science is *progress towards* anything,[5] while Weinberg thinks that science does progress towards something, that it 'brings us closer and closer to objective truth'. Weinberg quotes Kuhn: 'no sense can be made of the notion of reality as it has ordinarily functioned in the philosophy of science.'[6] Disputing Weinberg's idea of what progress involves is not the same as denying all progress.

This is Weinberg's final complaint:

> Kuhn's view of scientific progress would leave us with a mystery. Why does anyone bother? If one scientific theory is only better than another in its ability to solve problems that happen to be on our minds today, then why not save ourselves a lot of trouble by putting these problems out of our minds? We don't study elementary particles because they are intrinsically interesting like people – if you've seen one electron you've seen them all. What drives us onward in the work of science is precisely the sense that there are truths out there to be discovered, truths that once discovered will form a permanent part of human knowledge.[7]

Again, this trivializes Kuhn's deep reflections upon scientific progress and caricatures absurdly Kuhn's understanding of the scientific quest.

When confronted with the various views attributed to Kuhn, rather than thinking that this sounds either thrillingly provocative, or (as Weinberg does) dangerously misguided, one ought to pause for thought about what Kuhn means when he says these provocative-sounding things. Even if Kuhn says that scientific revolutions are like religious conversions, *just how* do they resemble them? What aspect of a religious conversion is Kuhn drawing attention to? Is religious conversion really the *very opposite* of 'an exercise of reason'? We should not either instantly react to Kuhn's claim in *The Road since 'Structure'* that *for his purposes*, the 'notion of reality as it has ordinarily functioned in the philosophy of science' (*TRSS*, 115) is of no use. Is it obvious what he is claiming? How *did* that notion 'ordinarily function' in the philosophy of science, and *just what* reservations does Kuhn have? We should not jump (as many do) to the conclusion that Kuhn is absurdly denying that there is any 'real world out there'. The critical element is that Kuhn is rejecting a doctrine in *philosophy of science*, and it would be quite wrong to infer from this that notions of 'reality' or 'nature' have no place in his account of scientific change.

Weinberg's is a standard critique of the legendary Kuhn who features as a bogeyman with which Realists and neo-Positivists terrify themselves and bully their opponents, and as a heroic adventurer to his admirers. Legendary figures often have little relationship to their actual human prototypes, and the legendary Thomas Kuhn has as little relationship to the real thinker as the legendary Robin Hood had to the person who occasioned his mythology. Robin Hood may not even have existed:[8] we have, however, a substantial basis in his own writings for identifying the real Thomas Kuhn.

The real Kuhn may prove a disappointment to the legend's admirers and critics alike. Depriving science of its 'objectivity' might seem a transforming achievement on the cultural landscape, revealing that the faith reposed in science is misplaced, and that science is 'no better than' astrology or witchcraft. Paul Feyerabend, Kuhn's 'anarchist' contemporary, raged against the cultural imperialism of science, but this critique is not one to which Kuhn gave any sign of assenting. In fact, views described in the manner Feyerabend described his own seemed to Kuhn 'vaguely obscene'.[9] Devastating the *philosophy of science* as Kuhn aspired to is no small achievement, though it need not – and for Kuhn did not – diminish the achievements of science. The critique of scientism – cultural imperialism in the name of science – can be undertaken quite independently of any views about the 'objectivity' or otherwise of science itself. Feyerabend's real attack was as much upon scientists as upon science, claiming that many scientists, and their cultural advocates, are ignorant and presumptuous (outside their technical expertise, where they persist in trespassing). They are ignorant of the practices and traditions that – 'in the name of science' – they seek to condemn, eradicate, or 'improve' and are presumptuous in thinking that their competence in a (nowadays, very restricted) domain of scientific work entitles them to – ironically – oracular status within the culture.

If Kuhn is allowed to be even modestly consistent – we present him as single-minded and dogged – he cannot seriously be considered as either a radical sceptic or a Relativist. Our defensiveness on Kuhn's behalf is not a product of uncritical devotion, and we identify important places where we judge that Kuhn's arguments get into trouble. Still, the first vital contribution to a critical assessment of Kuhn's views is to specify them accurately.

Historian and philosopher of science

Kuhn assiduously pursued two different, interrelated but, it transpires, unequal objectives, one historical, the other philosophical.

The first objective, when explained, may seem utterly innocuous, and now effectively achieved. Kuhn has turned out more successful than he anticipated, or finally approved of, in promoting a much more authentically historical history of science. His first objective was to do the history of science properly – to help *professionalize* the history of science[10] – primarily to make corrections to the history of scientific ideas. Kuhn's two substantial historical books (not counting *The Structure*) sought to reassess the achievement of crucial figures in the natural sciences: Copernicus in *The Copernican Revolution*, and Max Planck, often called 'founder of quantum physics', in *Black-Body Theory and the Quantum Discontinuity 1894–1912* (BB). However, it was what Kuhn made philosophically of his historical work that made his reputation.

The Copernican Revolution was a popular work, but studies of the quantum revolution in *Black-Body Theory* and an associated long paper (with John Heilbron) on Niels Bohr's model of the atom are relentlessly technical, no easy read for those without some mathematics and physics. Even though these studies carry Kuhn's general message, they would never gain him anything like the universal recognition that the racy and readable *Structure* won.

The professionalization of the history of science was meant to provide a platform for Kuhn's major interest in reconstructing the philosophy of science. The historical element was essential, not just because it would ensure the correct attribution of scientific achievements, but because it would enable an understanding of how scientists come to change their ideas, in a way that directly conflicted with the American philosophy of science Kuhn inherited and which he thought *basically mistaken*. Kuhn's 'case studies' were meant to show that the understanding *both* of what the succession of scientific ideas actually was, and of the ways those ideas changed was distorted.

Of Kuhn's two objectives, the historical is (at least, strategically and pedagogically) the prior one.

Given the importance of science in the history of the Western world, it is reasonable to suppose that the historical study of science ought to be an important part of history generally. However, when Kuhn first encountered the history of science he deemed it poor history indeed, done mainly by scientists, rather than historians, and, as such, principally intended for use in educating trainee scientists. As a result, the history seemed to Kuhn often misleading, suffering sorely from inappropriate 'benefit of hindsight'. The historical story often misrepresented the true record, inappropriately projecting recently developed concerns and understandings onto the scientific past. Even when historians rather than scientists did the history, it was likely to suffer the same fault, understanding the past too much in the terms of the present. Historical studies needed to be governed by the priority of understanding what science had actually been and how and why it changed, rather than either explaining or justifying our present scientific achievements to neophytes or the public.

Clearly, the proposal to approach the history of science from this direction is only worthwhile if it is expected that it will make a considerable and pervasive difference to the way the historical story is told. One of the main ways the history of science was distorted was that the nature of changes in specific scientific ideas was being misrepresented. Looking backwards, and viewing the thought of past scientists from the vantage of the present, one can be handicapped in understanding the actual nature of their thought. The big problem facing the history of science was much like that facing anthropology, that of understanding ways of thought very different from the researcher's own. As anthropologists have painfully learned, the solution is to keep alien thoughts in their proper context, not to isolate them from the considerations that make them intelligible and plausible to those who accept and live by them. *For historical purposes*, the important thing is not the relationship between the thought of some earlier scientist and contemporary scientific ideas. What matters is the relationship between that scientist's thought and the understandings available to the scientist in his or her own times, such as the thought of scientific contemporaries, the experimental data etc. then available, as well as influences from wider scientific and cultural traditions.

Kuhn was annoyed that scientific predecessors would, viewed with hindsight but in isolation from their historical context, be made out as inexplicably strange or confused: How could anyone 'in their right mind' possibly advance such strangely implausible ideas? How could earlier scientists (or, one might add, 'primitive people' in the anthropological case) fail to see the obvious, namely that they are *mistaken*? The only explanation would seem to be that they are less intelligent than ourselves: naïve, confused or incompetent. Or, 'at best', it might be questioned why, given that they were not all wrong, and did get part of the way towards where we now are, they could not see where their ideas would inevitably lead. What stopped them from seeing what is obvious to us?

Kuhn thought virtually all such judgements unjust or inappropriate, and expected that the re-examination of those thoughts in their full historical context would revise these negative evaluations. That a better understanding of the development of scientific thought will be gained by considering the different periods of science in their own right, rather than as mere stepping stones to the present, seems a modest and plausible suggestion. Yet, if true, it could once again have far-reaching implications, rewriting the history of scientific ideas themselves, and making a considerable substantive difference to the way in which the relationship between the changing ideas of science would be understood.

What do we mean when we say that the disapproved kinds of history of science *suffered* from 'benefit' of hindsight? 'Whig history' is a derogatory expression, condemning complacency in the writing of history. 'Whig' historians view history as though their present position and convictions are the very fulfilment of history, the outcome towards which previous periods have been inevitably leading. On this basis, the interest in previous periods is in the degree to which they progressed in the 'right' direction, namely towards the present state of things. Kuhn's reaction against this (and much more broadly than in the history of science) was to avoid the projection of present preoccupations into the past, seeking much more scrupulously for the contemporary understandings of the people of the period.

The attack on 'Whig history' is a questioning of what had been a ubiquitous notion, that of *progress*. If the notion of progress was not to be rejected altogether – and Kuhn never disputed that there was progress in science – then what was to be abandoned was the idea of 'progress' as the inherent, inexorable and predestined movement of history's own course (call this 'P'). It is a natural feature of science education, directed towards instilling up-to-date science in trainees, to look into the past mainly to see how the science got here from there. This history is an aid to education in current scientific ideas, rather than any end in itself. It *might* be acceptable as a part of basic scientific training, but cannot provide a satisfactory approach to real historical studies. The eschewal of Whig history correctively emphasized, then, the *contingent and non-special* nature of the present, making it merely the latest date in history, not its fulfilment or finale. The rejection of 'P' – of progress as involving the idea of movement to 'higher' or 'better' outcomes, and the complacent assumptions that one's own views were not only predestined but prove one's intellectual superiority over one's predecessors – is a rejection of only *one* idea of what progress might be.

That Kuhn denies there is progress in these terms does not mean that he denies *all possibilities of progress*, and one aspect of Kuhn's continuing struggle was to make clear the sense in which there is progress.

In Kuhn's view:

- External history (the history of the cultural and social circumstances in the wider environment of the working scientist) was in the service of internal history (recovering the course of the development of scientific ideas).
- Internal history remained in the service of *an* understanding of the past from the point of view of the present, providing a proper picture of how and why one scientific idea succeeded another.

Kuhn thought that the first problem confronting the history of science was correctly to identify *what it is that is to be explained.* It was here that the deficiencies of 'Whig history' were critical, for they produced mistaken accounts of past scientific work. Before one could explain how changes took place in science, one needed to identify those changes properly. The 'internal history' needed to be done in the right way, divesting it of retrospective distortion. 'Internal history' was also the main part of *the explanation*, explaining how, *through scientific work and reasoning*, radically new scientific ideas are proposed, tracing – as far as possible – the step by step transition from one set of ideas to the next. The way these things are *worked out* in the scientific documentation provides the main focus. An understanding of the transition as *a scientific transition* comes first, and only after that can 'external' history make its contribution, and show how 'external', non-scientific conceptions and interests (such as, for example, religious, political or commercial ones) might enter the scientific process. The influence of an area of science and of the wider society upon each other is a *bona fide* topic of inquiry, but, for Kuhn, the influence of the latter upon changes in the nature of the former can be understood only if the changes in the nature of scientific thought have already been properly understood. 'Internal history ' ensures this.

Thus, in *BB* – which is purely 'internal history' – Kuhn argued that the origin of the idea of a quantum discontinuity, a cornerstone of contemporary physics, was misunderstood by orthodox historical accounts. These located the break between classical and quantum physics more than half a decade earlier than was correct, and credited the wrong man, Max Planck, rather than its true originator, Albert Einstein.

Clearly, these 'main ideas' of Kuhn's about the history of science do not show how his work could ever seem a source of near-apocalyptic intellectual menace. But Kuhn was undoubtedly captivated by the philosophy of science and thought that a proper history of science implied profound changes in the philosophy of science. Kuhn's importance and influence come from his attempt to spell out the *philosophical* importance of the historical reformation. For the philosophy of science, Kuhn might indeed spell 'apocalypse'.

The history of science, as Kuhn conceived it, is bound to remain an esoteric pursuit. Involved in a close examination of highly technical scientific thought, it will not be understood by those without technical competence. What it shows will often be of little interest or consequence to most people, even working scientists in the area concerned. What difference could it make to most to know that the break between classical and quantum physics occurred after 1906 not 1900, or that Albert Einstein introduced the modern idea of the quantum, while Planck only introduced the term?

But there was a need to change the general (even popular) *image of science*. It was Kuhn's critique of this received 'image of science' (that is, the one dominant in the philosophy of science in the period around World War II) which made his work so controversial. The case for reconstructing the history of science was, for Kuhn, only a 'pretext' (though no trivial one) for demanding a thorough revision of the philosophers' image of (specifically) scientific change.

Why does a reasonably innocuous programme for the history of science carry such revisionary implications for the philosophy of science? Kuhn's main objective in the philosophy of science was to bring it into line with evidence about *what scientists actually do*, and this had the effect of altering the questions asked by the philosophy of science.

Kuhn did not mean that historical questions should displace philosophical ones, but that the idea of what was philosophically problematic should be drastically changed. It is this that explains some of the confusion which surrounds Kuhn's arguments, and which misleads those who persist in approaching Kuhn in terms of the more traditional philosophy of science standpoints.

Is scientific change a matter of continuous accumulation?

Kuhn's single-mindedness can be seen in the development throughout his major work, including the case studies, of one main theme: *that the received image of scientific development is of development by logically continuous accumulation, and that this is systematically misleading.* Insensitivity to intellectual and 'cultural' differences between different periods of science results in a failure to appreciate that differences between successive periods may be much deeper than they appear to cursory inspection; that the transition between one phase of science and its immediate successor does not so much involve continuing further along the same line of thought, as going off in a different direction. The cogency of the historical point remains: there is substantial and significant misrepresentation of the actual course of scientific work, partly out of a desire to get it to conform to the image encouraged by the philosophy of science. This misrepresentation is significant for what matters in philosophy of science, to wit, the status of science as knowledge and its standing as the very paradigm of rationality in (or indeed as the basis for the very rationality of) our general culture. In so far as the philosophy of science draws on the history of science, and forms its picture of the *bona fides* and rationality of science on that basis, *then if the history is seriously misguided, so too will be the philosophy 'modelled' on it.*

If we treat Kuhn's work as focused single-mindedly on devastating the view of science as development by (smooth, continuous) accumulation, no serious injustice will be done to the range and variety of his writings, for that view is ramified, and Kuhn's criticism many-sided. It will facilitate our account of Kuhn's confrontation with development-by-accumulation if it is borne in mind that Kuhn's attack is on views in the philosophy of science, and not on science itself. Various positions in philosophy (though usually to the disadvantage of reasoned discussion with them) profit by identifying themselves with science (wrapping themselves in its flag), to the extent that they make it seem that rejection of their views diminishes the standing of science itself.[11] Thus, if one argues against *their view of* what makes science rational, one may be accused of denying that *science* is rational. Similarly, questioning their idea of the nature of scientific knowledge will be responded to as 'radical scepticism' doubting whether science can truly attain knowledge.

Those with more traditional views in the philosophy of science are sometimes prone to overreaction – some, certainly, are easily outraged. However, there is no reason to be intimidated by a fierce insistence that science loses all credibility if their philosophy is rejected.

Kuhn argues that if we go along with the development-by-accumulation image, then we also subscribe to received conceptions of rationality, scientific progress and the growth of knowledge, and of the requirements for knowledge of natural reality, *even though these notions involve serious misconceptions about what scientific change has really been like.*

Kuhn's arguments tell not against the achievements of science, nor against its prospects for rationality, progress, growth and the accumulation of knowledge, only against one rendition of these concepts, the received (in practice *ahistorical*, and mostly formalistic) philosophy of science.

What exactly is the image that Kuhn proposes to give up? We have (following Suppe)[12] dubbed it the 'received view'. (We are not trying to give an account of this received view as it might be expounded by one of its best supporters, but one of the enemy as *perceived* by Kuhn, but this does accurately reflect substantial elements of the philosophy of science of Logical Positivism *and its successors*, and also important elements of the views of Falsificationists and of 'Realists'. The 'received view' is probably what many lay people, especially those informed by 'popular science' or popular philosophy, still understand as science-in-itself.)

The trouble with development-by-accumulation as outlined so far is that it is neither clear nor specific, and needs to be spelled out. Kuhn's claim is that science does not develop *only* by accumulation, and that it does not develop by *constant* or *linear* or *logically smooth* accumulation. If this is what Kuhn is arguing, then the perceived opposition to him must be maintaining that science develops by constant, continuing, linear accumulation.

Ian Hacking picks upon Carnap as a representative Verificationist, and then 'opposes' his views to Popper's, in a manner clear and helpful to us:

> Carnap thought that *meanings* and a theory of *language* matter to the philosophy of science. Popper despised them as scholastic. Carnap favoured *verification* to distinguish science from non-science. Popper urged *falsification*. Carnap tried to explicate good reason in terms of a theory of *confirmation*; Popper held that rationality consists in

method. Carnap thought that knowledge has *foundations*; Popper urged that there are no foundations and that all our knowledge is *fallible*. Carnap believed in *induction*; Popper held that there is no logic except *deduction*.

All this makes it look as though there were no standard 'image' of science ['received view'] in the decade before Kuhn wrote. On the contrary, whenever we find two philosophers who line up exactly opposite on a series of half a dozen points, we know that in fact they agree about almost everything else.[13]

Carnap and Popper share a basic image of science, that rejected by Kuhn.

What then *is* the common ground between the two? Hacking again:

> Both think there is a pretty sharp distinction between *observation* and *theory*. Both think that the growth of knowledge is by and large *cumulative*. Popper may be on the lookout for refutations, but he thinks of science as . . . tending towards the one true theory of the universe. Both think that science has a pretty tight *deductive structure*. Both held that scientific terminology is or ought to be rather *precise*. Both believed in the *unity of science* . . .
>
> Both agreed that there is a fundamental difference between the *context of justification* and the *context of discovery*. . . . Philosophers care about justification, logic, reason, soundness, methodology. The historical circumstances of discovery, the psychological quirks, the social interactions . . . are no professional concern of Popper or Carnap. . . . [T]he philosophies of Carnap and Popper are timeless: outside time, outside history.[14]

Kuhn can be seen as disagreeing with all the Popper–Carnap common ground. His antagonism is manifest in the list of denials that Hacking attributes to Kuhn:

> There is no sharp distinction between observation and theory.
> Science is not cumulative.
> A live science does not have a tight deductive structure.
> Living scientific concepts are not particularly precise.
> Methodological unity of science is false: there are lots of disconnected tools used for various kinds of inquiry.
> The sciences themselves are disunited. They are composed of a large number of only loosely overlapping little disciplines many of which in the course of time cannot even comprehend each other . . .

The context of justification cannot be separated from the context of
   discovery.
Science is in time, and is essentially historical.[15]

One way of construing the development of natural science would
be to see science as just the steady addition of new scientific results
to a vast stockpiling of results. Kuhn rejects this, since it doesn't take
much thought to appreciate that the development of science is not
just producing new findings stacked on the old in no particular
order. An important part of science is putting findings into an orga-
nized form, especially connecting them together through systems
of theory. Many major achievements in science are other than
producing empirical results, and involve contriving new ways of
thinking about problems and new ways of investigating. Hence, if
science develops, it does not do so just by further addition but also
by rearrangement and perhaps subtraction. More importantly, in
the development of science, especially in the formation of major theories, new theories *displace* old ones. Everyone knows that Einstein's
theories have displaced Newton's, even if only in that Newton's
theory has been swallowed up into Einstein's.[16] There *are* significant
changes in science, sometimes great 'revolutions', but these are – on
the views Kuhn contests – not fundamental breaks in the nature
of scientific thought but 'great leaps forward' that carry science
continuously further in the same general direction. This is what is
meant by the idea of development as 'linear' – that the advance of
science does come about through substantial discontinuities
between ways of thinking, but through logically continuous exten-
sions, taking science nearer to its ultimate objective. This gives one
a way of seeing scientific change as truly a case of *development*.
   So much has this been taken for granted that we have scarcely
even explicitly mentioned it yet – the idea that change in science is
(overwhelmingly) some form of *improvement* (that is, development
in a positive sense). Science gets better, science just *knows* more. In
some sense, this idea surely demands to be accepted.[17] Does science,
though, just go on adding more and more findings to a stockpile
that can expand endlessly? Or is there some 'end' or 'aim' for
science? One positive answer given to this question is that science
is advancing towards the full and final truth, getting ever closer, if
not (as Karl Popper doubted) ever actually getting there. Over time
it approximates more closely to this final target. Succeeding ideas
are each measured against the same target, and the latest ideas are
accepted when they are nearer to, or even hit, that target.

Thus arises another element involved in a conception of the continuity of science: of the move from one stage to another as a *rational* progress. There is – perhaps – constancy throughout the development of science in the means scientists use to decide whether to accept a new idea, which involve applying the same general criteria. Whether or not to take a particular step in science is thought of as a premier instance of applying *reason*. Scientists do not decide on next steps through violent physical struggle, or drawing lots, or conducting opinion polls. They decide by thinking through – logically reasoning and debating – the differences between old and new. A strong influence in philosophy is the idea that reason, if done properly, follows rules. There must be universal rules that comprise the application of reason, and that can be spelled out. This is much the same as saying that knowledge is a product of method, and that if science is the paradigm of knowledge, unparalleled in the systematic accumulation of knowledge, there must be some 'scientific method' that is followed in all successful cases. *If* there is such a method, then, once identified, it provides a sure-fire method of getting the right results, of ensuring that *knowledge* is obtained, a method that can be extended to all those areas it has not yet been applied to. Further, if thinking is done 'rationally', all rational beings will be logically compelled to accept its conclusions: here again is the 'universalism' often attributed to scientific method.

The above is a much simplified account and the positions included in this compact package are often more complex and sophisticated than this bare summation depicts. We are only trying to show how Kuhn's critique of development-by-accumulation could wreak havoc in the philosophy of science, for it questions *all* these contentions. He claims that these are not – that they cannot be – descriptions of how scientists actually choose to move on or stay put. Indeed, it is significantly misleading even to say that scientists are involved in *choosing* between one scientific idea and another.

An important role many see for the 'received view' is that it reassures us. It provides us with the assurance that our contemporary science truly is the best account of natural reality, that it is right to repose in it the high level of trust that many do. There is no equivalent or better account of natural reality to be found in either previous historical periods or in any of the other societies currently in existence. The 'received view' apparently gives us such assurance for it tells us both:

(a)   that there is progress in science itself, that the latest science is the best and, of course, is more advanced than prior science, and than all pre-scientific modes of thought; and

(b)   that because natural reality has a determinate character, then any characterization which truly describes, as science does, how natural reality is can have no competitor. Natural reality can be one way only.[18] Since our natural science tells us how natural reality is, our science can have no rival.

These reassurances are deemed hugely important. Given the great weight placed upon science in our cultural self-conceptions, they justify our thinking of ourselves as advanced and enlightened people (not because we ourselves contribute to science, but because we are the – fortunate – inhabitants of a culture pervaded by and properly respectful of the scientific attitude). Naturally, for anyone holding such views, it is incumbent on them to defend science against its critics – they think that far-reaching and deleterious consequences follow science's demotion.

These latter convictions have lately felt embattled, facing a rising tide of 'irrationalism' threatening science and civilzation, and their defence is all the more vigorous because of the seriousness of the threat perceived.

The science wars

Scepticism about such convictions is an important part of 'postmodernism', and postmodernism is seen as a resurgence of 'irrationalism', to be feared because its previous prominent manifestations have allegedly been the totalitarian extremisms of Nazi Germany and Soviet Communism. Thomas Kuhn is readily and regularly blamed for his major role in this (supposed) contemporary irrationalist resurgence and therefore labelled as a serious political danger.[19]

Kuhn unquestionably tells us that science is not what we have believed it to be (*in so far as we have believed the philosophy of science and its popularizations*). If we accepted science as the sole unqualified paradigm of rationality, and its results as uniquely successful in telling us about natural reality, then the easiest conclusions to draw would be that Kuhn undercuts this image by promoting its very opposite. In that case, Kuhn must be telling us:

- that science does not tell us about the nature of natural reality at all; or
- that science is not uniquely successful in telling us about natural reality, and that other, very different, ways of talking about natural reality are (or are capable of being) as successful as our natural science.

To those holding the received view such suggestions seem not only dangerous, but preposterous, adding to the accusatory and intemperate tone often taken towards Kuhn. He is blamed for undermining our faith in science (and in our enlightened way of life). And it is for precisely this reason that some of his more 'radical' postmodernist 'fans' think him worth following.

What of 'foundationalism' in philosophy? Is Kuhn an 'anti-foundationalist'? And if he is, could that too make him dangerous? 'Anti-foundationalism' is another buzz word, identifying opposition to the classic notion that knowledge must be 'founded' on absolute certainty. That view is now in relatively bad repute, and Kuhn is hardly alone or very distinctive among distinguished philosophers of science in eschewing it (Popper too rejects foundationalism). Kuhn rejects the traditional, empiricist source of certainty, namely, direct access to our sensory experiences that provide, after all, the most immediate and uncontaminated contact with the external world. In the empiricist philosophy of science, this idea took the form of a distinction between two languages, one made up of 'theoretical terms' and another consisting of 'observation terms' (SSR, 113–29). The latter speak directly of our sensory experiences, and therefore characterize those experiences independently of theoretical terms and identically from one observer to another (assuming that the immediate, uninterpreted sensory input would be the same for all of them). This was a putative basis for the rationality and objectivity of scientific choices. *If the observation language is constant in the face of the diversity – and rivalry – of theories, it is something by means of which those rival theories could be unequivocally compared.* One could tell whether or not what was said by one theory did, as its rival did not, correspond with what the observation language could report as observed with certainty. Kuhn – though not original in this[20] – questioned the distinction between observational and theoretical terms,[21] denying the existence of *theory neutral* observation reports. Kuhn eschews the observation language/theory language distinction *and does say* that science *never* involves comparison of a theory with nature itself. Is this another 'dangerous' view, risking the thought that the world is only a world-for-us, a world dependent upon being perceived? That would be philosophical Idealism.

Kuhn *is*, to be sure, anti-foundationalist, and in yet another way. Those holding the received view think that they are providing *a philosophical* foundation for natural science, securing the claim of science to be rational, objective and cumulative. Kuhn effectively denies science such external philosophical justification, denying that there can be any external philosophical basis for judging the success of science. Kuhn thus disposes of the idea that philosophy of science is important, either justifying or second-guessing scientific results. No wonder some philosophers of science are enraged.

Kuhn dissents, and in almost every particular, from the 'received view', but does *not* simply negate claims that science is rational, cumulative, and accesses natural reality. Rather, he challenges the terms of his opponents' arguments, refusing to conceive rationality, cumulation and knowledge in the sciences in the way the received view does. If Kuhn is right, that the historical facts about scientific change just do not conform to the 'image' projected by the received view, it does not follow that one must suppose that science is irrational: one can argue that to the (great) extent that science *is* rational, *it is so in a way very different from the received view's ideas.*

Thus, Kuhn has been prominently criticized – and sometimes praised – for

- Irrationalism;
- Relativism;
- Idealism.

Much opposition has been premised on finding him guilty of one or more of these offences. However, Kuhn was none of the above, except perhaps in the mildest and most unobjectionable fashion. Kuhn is only seriously open to construal as an irrationalist, Relativist or Idealist if one accepts the received view as one's starting point, and construes Kuhn's writings as if they were designed to satisfy the requirements of that view. But *that view is what is in dispute.*

Kuhn's 'schematic' of the development of natural science

Kuhn's *The Structure of Scientific Revolutions* is a sketch[22] of how to think about the general development of *natural sciences* in the West.

Before our chapter-by-chapter exegesis of *SSR*, let us outline Kuhn's 'schematic'.

Disciplines that have become developed or mature sciences – and there is no necessity that any discipline that exists must end up in that collection – are, in their early life, made up of thinkers and investigators whose work is only tenuously, often competitively, connected. Often, these thinkers divide into subgroupings ('schools') that are rivals in attempting to elect themselves as the group defining the whole discipline for everyone, but they are ineffectual, and no one of them can mark itself out as sufficiently successful to attract more than a proportion of those working in the area. The discipline fails to make anything the participants can recognize as real progress, and the result is frequently a feeling that it is going about things in basically the wrong way, and that a thorough rethink of the whole discipline is needed. While empirical work may get done, much of what goes on involves disputing the fundamentals of the discipline, and as long as those are constantly being reworked then work is not going to move on from first base.

Out of this factional strife among these schools, there may appear some achievement sufficiently attractive to a large proportion of the participants in the field for one to begin to talk about the discipline as unified. (Fundamentally, such a novel achievement is *what Kuhn is talking about when he uses the word 'paradigm'*, probably the most famous of all his expressions.) This does not mean that everyone has gone over to this new position, but the balance has shifted, and many of those in the field can conduct themselves on the basis of *assuming agreement on fundamentals with most of their colleagues*, and can give up either reiterating or arguing about those fundamentals. There may remain a minority who still contest the fundamentals, but these will eventually be 'frozen out' by the others. They will not necessarily stop their dissent, but will just be disregarded. Those who have opted in to the now prevailing agreement take each other seriously, but cease to take any interest in the work of those refusing to sign up. Where the discipline was previously open to all with scholarship in the field, the fundamental agreement brings a more exclusive attitude, and the development of organization in a more professional direction, for instance with an internal system of control, including monopoly over the training that becomes essential to recognition as a *bona fide* practitioner.

Before the development of this extensive agreement on fundamentals, the discipline could be described as being one without a 'paradigm'.

Once a field has developed this paradigm, the consequences of settled fundamentals come into play. Work within the field can start to build on the fundamentals, and still other work can build upon that. There will be a (relatively) standard sense of what an achievement in the field looks like, and a (relatively) coherent body of knowledge articulated, to which the results of appropriately conducted studies count as additions. Thus knowledge-by-accumulation becomes possible – and this, when it happens, is surely a fine thing.

A paradigm, facilitating such cumulation, will not however remain permanently in place. Much of what goes on in the discipline will be incorporated into the paradigm, but there will be things that do not fit within it – results that do not conform to what the principles and orthodox practices of the science would lead one to expect. At some point, the incongruity between those results and their paradigm will encourage doubt as to whether there is not some flaw in the fundamentals, and whether they need – in whole or in a major part – to be rethought. If those incongruities – call them anomalies – are seen as raising serious questions about the viability of those fundamentals, then the discipline will exhibit again some of the behaviour it showed in the period before the fundamentals were settled, and will involve, again, internal division. However, the extent to which this happens will be more limited than in the earlier days before agreement, for even though those in the field are prepared to wonder about, even re-examine, those fundamentals, they will not abandon them until they see some alternative fundamentals that have the prospect of replacing those in question. They do not regress to the stage in which the discipline collapses back into rival schools, but move into a period of 'revolution' in which there will be division between those who want to stick with the old fundamentals, and those who want to follow the new way. From their competition a new and prevailing consensus will emerge, a group will break off and form a new discipline, and a new and different paradigm will come to dominate – as before, marginalizing those who do not sign up. After a transitional period of some disorganization – even of 'revolutionary struggle' for control – the discipline will reform itself in a reorganized way. Much of what the discipline had achieved before will be rethought, rearranged or forgotten, but, more importantly, the discipline develops new possibilities that could not have been imagined or taken seriously in it previously. Now, on the basis of radically revised fundamentals, a new phase of building is possible (unless and until, of course, the possibility of serious doubt about the viability of the latest set of fundamentals arises).

Here is Hacking's account of the situation, after a revolution in a science: '[Any] new theory . . . is born refuted. A new generation of workers gets down to its anomalies. There is a new normal science. Off we go again, puzzle-solving, making applications, articulating mathematics, elaborating experimental phenomena, measuring.'[23] And as Hacking observes, and as we have noted, the process of change in fundamentals will in all likelihood sometime begin again: 'Kuhn invites the idea that every normal science has the seeds of its own destruction.'[24] For if a paradigmatic theory is born refuted, as Kuhn thinks, then there will always be room for working on it – and always scope for wanting or 'needing' to overthrow it.

The pattern Kuhn (and Hacking) describe of 'dialectical movement' in the natural sciences has brought about a broader pattern in which there is long-term advance for some disciplines, and general expansion in the range of disciplines, with increased levels of specialization within and among them. The possibility of 'long-term advance' is present in the business of paradigm succession, for though the change of paradigm results in rejection of the previous paradigm as such, this never involves going back on *all* its achievements, for many are reincorporated within the new paradigm, *albeit typically in radically reconsidered terms* (as we clarify later, this is a central source of Kuhn's philosophical radicalism). The new paradigm achieves acclaim by raising the level of scientific performance, being acknowledged as generating some, *and promising many more*, achievements that do not merely equal but surpass (in terms not merely of finding out new things, but of doing so in more technically demanding ways) those of the previous paradigm.[25]

Kuhn's historical-cum-sociological sketch of the development of natural science lays the groundwork for his philosophical innovations. Those are what we shall endeavour to describe and assess in the pages which follow. We wish to explicate Kuhn's 'schematic', and show thereby *both* its reasonableness *and* its revolutionary effects.

Part I

Exposition

We now expound Kuhn's three major works, *The Copernican Revolution* (CR), *The Structure of Scientific Revolutions* (SSR), and *Black-Body Theory and the Quantum Discontinuity, 1894–1912* (BB). Hardly anyone, even authors giving lengthy accounts of Kuhn's work, read his historical studies in conjunction with *SSR*, and this is, we think, one reason why the understanding of Kuhn is characteristically so poor. *SSR* is packed with historical examples, but these are brief and illustrate specific points of argument, giving insufficient guidance to the difference Kuhn's ideas were meant to make to our picture of science. The failure to understand how the ideas in *SSR* cash out in the historical studies almost invariably signals a failure to understand both. Exposition of Kuhn's main case studies is, then, in practice, an essential aid to explaining *what it is that Kuhn is talking about in SSR*. Bringing these case studies into our discussion more heavily than other authors have done will make it easier to see that Kuhn's arguments cannot be as *absurd* as they are accused of being. Much or all of what Kuhn says about the Copernican and quantum cases may (as a matter of historical claim) be partly or wholly false, but the kinds of claims he makes are intelligible enough, and propose nothing bizarre or fantastical. Kuhn's claims are, it is worth remembering throughout, claims about the working practices of natural scientists and *only* about that.

1

The Structure of Scientific Revolutions

Our presentation of *SSR* will itself be in two segments. We explicate the main elements of Kuhn's account of the dynamics of natural science in the West, paying particular attention to those that have been provocative or have spawned confusion. Then we introduce some of the philosophical 'matters arising' (that are dealt with more critically in part II) from that sketch of scientific change. Our account is *roughly* correlated with sections in *SSR*. (We deal with Kuhn's amendments to the main body of *SSR* later.) In the first segment of this chapter we deal with Kuhn's central – and, we think, fairly straightforward – concepts for depicting the main changes that take place in natural science: 'paradigm', 'normal science', and 'scientific revolution' provide key words. In the later segment we deal with ostensibly more problematical parts of Kuhn's case, where he introduces (initially at least) very strange sounding ideas: those of 'world changes', 'phenomenal worlds', 'incommensurability', not to mention his rejection of the idea of 'a fixed nature'. We will make a first attempt to show that these ideas are not as strange as they can seem, although that is a long way from maintaining that they are free of difficulties. Whether the difficulties basically invalidate Kuhn's approach is dealt with in part II of the book.

I

If Kuhn's image of science is as 'innocuous' as we say, how is it possible for his work to carry the extreme implications that are regularly attributed to it? We have already suggested that this is less because of what he says than because of what, in saying what he does, he undermines: various popular and ingrained and academically 'respectable' views about the sciences. He thoroughly undermines *not the sciences*, but entrenched *philosophical* assumptions *about* them. And that is not felt, by those he is subverting, to be innocuous.

But what does he actually *say*? And what does he mean by it?

His central platform is set out in *The Structure of Scientific Revolutions*. With some *comparatively* minor modifications, this is the one to which he adhered in the rest of the work published in his lifetime.

In the Preface to *SSR*, Kuhn writes the following autobiographical note, which we take to be absolutely central and essential to understanding his project:

> I was struck by the number and extent of the overt disagreements between social scientists about the nature of legitimate scientific problems and methods. Both history and acquaintance made me doubt that practitioners of the natural sciences possess firmer or more permanent answers to such questions than their colleagues in social science. Yet, somehow, the practice of astronomy, physics, chemistry and biology normally fails to evoke the controversies over fundamentals that today often seem endemic among, say, psychologists or sociologists. (*SSR*, viii)

Thus, the difference so important to Kuhn's thought is one which he first notes in connection with *contemporary* work and the division between natural and social sciences, and that he realizes can be projected back into the history of the natural sciences themselves.

Further down the same page, Kuhn adds that *SSR* is 'an essay rather than the full-scale book my subject will ultimately demand'. Kuhn never wrote that book. A first step towards 'constructing' it is to gain a thorough base-level understanding of *SSR*. When that is combined with Kuhn's later (and earlier) work, a thorough picture of how that book might at least be virtually constructed is possible.

A work of history or of philosophy of science?
(On section I of *SSR*, 'A role for history')

Let us begin by mentioning again the central issue of the succession of major theories in science. Kuhn takes the received view in the philosophy of science to be making big claims about what criteria are *and should be* used to choose one scientific theory over another, and argues that these claims are 'falsified' by the historical record. Thus, a main purpose of *The Structure of Scientific Revolutions* is to make a case as to how scientists *do in fact* come to replace one theory with another. This makes it sound as though *SSR* is one of his historical studies, but it is not that. How, then, is the historical stuff a 'stalking horse' for the philosophical; how is this latter aspect dominant in *SSR*? *SSR* differs from Kuhn's properly historical studies for he is not, here, primarily concerned to detail what, as a matter of historical fact, occurred in various specific episodes in the history of science, but, instead, to say how the events in such episodes should be philosophically construed.

In the Introduction to *SSR*, Kuhn noted that he was already implicitly or explicitly querying such verities as, for example,

> the very influential contemporary distinction between 'the context of discovery' and 'the context of justification' . . . [H]aving been weaned intellectually on these distinctions . . . I could scarcely be more aware of their import and force . . . Yet my attempts to apply them . . . to the actual situations in which knowledge is gained, accepted and assimilated have made them seem extraordinarily problematic. Rather than being elementary logical or methodological distinctions, they now seem *integral parts of a traditional set of substantive answers to the very questions upon which they have been deployed.* (8–9; emphasis added)

He ends the Introduction with the following, ringing question, which seeks rhetorically to insinuate that the needed transformation of philosophy of science has begun: 'How could history of science fail to be a source of phenomena to which theories about knowledge may legitimately be asked to apply?'

From immature to mature science
(On section II of *SSR*, 'The route to normal science')

It is important to emphasize that Kuhn is *mainly* concerned in *SSR* with the revolutionary transformation(s) of *mature sciences*, and not with the initial transition from immature to mature: the latter is only a preliminary, though important concern, highlighting the difference between a situation in which no cumulation of knowledge takes place . . . and one in which it persistently does, albeit with disruptions.

When Kuhn remarks (10) that 'normal' science is what most scientists spend most of their time doing, he is talking about operating within a setting where there is 'agreement on fundamentals' already in place, the possession of such agreement on fundamentals being the hallmark of a mature science. In the first instance, the notions of 'paradigm' and 'normal science' are not meant to express the difference of 'normal' science from 'extraordinary – i.e. revolutionary – science' but to capture, rather, *the difference between sciences that do and do not have* this kind of fundamental agreement. The *first* contrast that needs to be kept in mind, then, is between

- (a) those areas of study in which a good many things are settled, and where there is some kind of broad consensus on the nature, main business and prevailing approaches of the enterprise; and
- (b) those pursuits in which there is little if anything settled.

The contrast between physics and sociology over the past three hundred years is a good example of what Kuhn has in mind. There have been 'revolutionary' upheavals in physics but between these revolutions there have been stable and extensively shared frames of reference that encompass the vast majority of physicists. The same cannot be said of sociology, for example, which, though two hundred and more years old, is far from attaining anything approaching unification. The notion of 'paradigm' is meant, then, to serve in the first instance to illustrate (or constitute) the contrast between a science like post-seventeenth century physics and a would-be science like sociology (*SSR*, 15).

The great difference from the point of view of the practice of science is, for Kuhn, that it is only when there is extensive agreement among them, in their suppositions and practice, that scientists can really get on full-time with the job of empirical research, rather than being constantly diverted from this by the need to argue about the justification and rationale of what they do.

In this context, the notion of 'paradigm' functions actually in at least two distinct ways – and in his 1969 Postscript to *SSR* (in the second and third editions), Kuhn accepted that he had not demarcated the two as clearly as he might. (He would later adopt 'disciplinary matrix' as a more univocal term than 'paradigm' for referring to the encompassing and extensively (though never utterly) uniform body of assumptions shared within a mature discipline.[1] He attempted to restrict the word 'paradigm' to the paradigms – exemplary achievements – which founded sciences and around which subsequent revolutions were built (via their function as models of scientific 'good practice').)[2]

Kuhn argues that many of the natural sciences started off being more like sociology than like physics after Newton. Sciences often begin with a phase in which they are like the contemporary social sciences, where there is no fundamental agreement, where people keep trying to rebuild the science all over again, tearing up existing views of its nature and purpose, and trying to make a completely fresh start. The natural sciences that we now have, and that did start off that way, at some point decisively left this pre-agreement (pre-paradigmatic, as it is sometimes called) state behind them, never to return to it. Kuhn recounts the successive changes in the view taken by physics of the nature of light since Newton: light was conceived as corpuscular, then as waves and then as photons. Each was, in its turn, and for a time, the generally accepted view within physics. However, until Newton, 'no period between remote antiquity and the seventeenth century exhibited a single generally accepted view about the nature of light. Instead there were a number of competing schools and subschools' (12). So, the contrast is between the period prior to Newton, before the seventeenth century, when there was no general agreement in optics as to the nature of the phenomenon of light, but only competing views, and the period after Newton when there were drastic changes in the conception of the nature of light, but, *at any one time*, pretty general agreement held on a current view.

Kuhn holds that the predecessors of physics are rightly considered scientists,[3] and they 'made significant contributions of the body of concepts, phenomena and techniques from which Newton drew the first nearly uniformly accepted paradigm for physical optics' (13). But, while 'these men were scientists', anyone

> examining a survey of physical optics before Newton may well conclude that, though the field's practitioners were generally scientists, the net result of their activity was something less than science. Being able to take no common body of belief for granted, each writer on physical optics felt forced to build his field anew from its foundations. (13)

Not much serious cumulation in such a case! Clearly, the force of
the contrast between the ventures with and without paradigms
highlights the difference between the case

(a)   in which there is, proportionately speaking, relatively little
      real scientific work (in the sense of empirical *investigation*
      etc.) and where that work is not in any real sense cumula-
      tive but is, rather, randomly assorted, being undirected
      and uncoordinated, with no integration between results of
      different studies, between the work of one scientist and
      another in the same field, and with no possibility of build-
      ing further investigations upon established techniques and
      accepted findings; and
(b)   in which there is unity and coherence in the investigations
      carried out, with generally accepted ideas and procedures
      and where the findings of one study build directly on those
      made by another.

Development-by-accumulation, again

Kuhn is a severe critic of the image of development-by-
accumulation, but, as suggested earlier, he does accept *to a significant extent* the picture of science as involving the stockpiling of
knowledge. *However* he does so on the basis:

(a)   that the cumulation takes place against the background of
      a considerable measure of agreement on fundamentals,
      and on the basis of treating certain past achievements as
      generally exemplary, as a guide to how to do further work,
      providing a framework within which meaningful accumu-
      lation is possible, and a context in which each scientist no
      longer needs to be involved in beginning all over again for
      themselves;
(b)   that the paradigm is seen to have been a crucially missing
      element from the simple stereotypical ('received') picture of
      development-by-accumulation, but is in practice *taken-
      for-granted in that picture*, as it is in normal scientific work
      itself; and
(c)   that occasionally paradigms are overthrown and the 'devel-
      opment by accumulation' has to begin again, from a dif-
      ferent – but 'upgraded' – starting point.

While a scientific field may develop out of a 'pre-paradigmatic' phase, once it has developed a paradigm, thereafter new disciplines and specialisms may spin off from that area of work without themselves having a prior pre-paradigmatic phase.

The emergence of a paradigm shifts the situation within the field, creating a more rigid and exclusionary situation; some people go along with the new paradigm, others literally die without having been able to reconcile themselves to the change.[4] The group associated with the now dominant paradigm is transformed from a loose group with shared scholarly interests into a profession with all its appurtenances, with journals, specialist societies and a control over qualification in its field. Textbooks, not the working scientist, take on the job of spelling out the science's fundamentals. Advanced research becomes interesting and accessible only to specialized colleagues and the scientific paper, rather than the book, becomes the means of communication within the profession (20).

The importance of *organization* now becomes very clear. It is not as if other philosophers of science had denied the existence of this *professionalizing* tendency of science, they just didn't seem very interested in it. Even common sense might have noticed it since, as the following quotation avers, 'it has become customary to deplore' this development, and to regret 'the widening gulf that separates the professional scientist from his colleagues in other fields, [though for Kuhn] too little attention is paid to the essential relationship between that gulf and the mechanisms intrinsic to scientific advance' (21).

The concept of 'paradigm'

Kuhn's most famous concept[5] is that of 'paradigm'. Not one that he initially coined, it has however caught on since he adopted it. One hears almost endlessly nowadays of 'new paradigms' arising or being needed in every area from Geology or Child Psychology or Management Science to the 'New Age'. Of course, just this ubiquity should be a cause of concern for us. What concept could it possibly be that could be understood and serve so widely?

Naturally enough, it turns out on closer examination barely to be Kuhn's concept at all. For the first point that must be borne in mind here is that, even in Kuhn's own work, the term 'paradigm' stands *for very different things*.

Let us begin by looking at where Kuhn begins, at the point where the term 'paradigm' – in Kuhn's particular sense(s) of it, in its paradigmatic sense for him – gets introduced:

> In this essay, 'normal science' means research firmly based upon one or more past scientific achievements, achievements that some particular scientific community acknowledges for a time as supplying the foundation for its further practice. Today such achievements are recounted by science textbooks. Before such books became popular many of the famous classics of science fulfilled a similar function. Ptolemy's *Almagest*, Newton's *Principia* and *Opticks*, Lavoisier's *Chemistry* – these and many other works shared two essential characteristics. Their achievement was sufficiently unprecedented to attract an enduring group away from competing modes of scientific activity. Simultaneously, it was sufficiently open-ended to leave all sorts of problems for the redefined group of practitioners to solve.
>
> Achievements that share these two characteristics I shall henceforth refer to as 'paradigms,' a term that relates closely to 'normal science'. By choosing it, I mean to suggest that some accepted examples of actual scientific practice – examples which include law, theory, application, and instrumentation together – provide models from which spring coherent traditions of scientific research. These are the traditions which the historian describes under such rubrics as 'Ptolemaic astronomy' (or 'Copernican'), 'Aristotelian dynamics' (or 'Newtonian'), 'corpuscular optics' (or 'wave optics'), and so on. (10)

It is interesting to note how strongly this, Kuhn's initial account of paradigms, draws on the 'literary' aspect of science: Kuhn emphasizes the importance of classics (then) and textbooks (now) – virtually no natural scientists read the classics any more (having no need to) in defining their field(s). Textbooks, written works, play a major role in laying down what a paradigmatic scientific achievement is, how it is to be understood, how it is to be *taken and used*.

Also emphasized is the sharedness and indeed *compulsoriness* of the paradigmatic. This is already a strong hint that any suggestion that one can choose to have or even try to have a scientific revolution, to move to another paradigm, is going to be wrong-headed. One is enormously constrained – by the world in one's lab . . . and by one's tradition and community.

Let us now focus in on what was for Kuhn the heart of his conception of 'paradigm', the sense in which the word must sometimes be meant (if one is to have an effective philosophy of science), whatever other senses it might also be used in.

Exemplars (artefact paradigms/construct paradigms) This is para-digms as exemplary, as acknowledged achievements providing models to follow, laws to explore and to find new versions of in new sets of circumstances, *etc.*

This usage of the term 'paradigm' derives from teaching grammar. A paradigm in grammar is literally an example that one is supposed to be able – once one has understood it – analogically to apply in new circumstances. For (a very simple) example: if one is given the endings to a verb in French (such as *bouger*), and then told that these are the endings to all such verbs – to all verbs ending in '-er' – then one has a paradigm for conjugating those verbs oneself. Similarly, Kuhn thought, in science – with an important proviso: in grammar

> the paradigm permits the replication of examples any one of which could in principle serve to replace it. . . . In a science, on the other hand, a paradigm is rarely an object for replication. Instead, like an accepted judicial decision in the common law, it is an object for further articulation and specification under new or more stringent conditions. (23)

Just as with the disciplinary matrix, so with the exemplars which form a vital part of the former, the extent to which they actually are fully-mutually-understood is, according to Kuhn, uncertain prior to being tested in times of crisis. It may turn out that they always were understood differently by different scientists, but the difference never previously gave rise to an issue as relevant cases never arose – the shared understanding and application of the paradigm, of the exemplar, never *needed* to be in question.

This is perhaps because rules of scientific procedure are rarely explicitly taught, but are rather absorbed with the paradigms – and the way in which the paradigms are presented and instilled may vary a little from one educational setting to another, resulting in variable understandings as to how exactly to go on from the para-digm. By hypothesis, being always potentially on the cutting edge of research, it must have unapplied instances easily within reach, and one must be ready for scientists to find that they do not agree quite as much as they thought they did, that when it comes to this new case, they diverge in their judgements of what the right thing to do is (though whether or not this will matter or ever be noticed will depend on circumstances).

Kuhn's use of the term paradigm can seem quite unsatisfactory in its ambiguity – though certainly not as unsatisfactory as Margaret

Masterman has led people to believe[6] – but it would help, perhaps, to notice the (harmless) *crudity* of the exercises that we are involved in here. Kuhn's account of the development of science is a pretty *gross* one. It is no more *gross* than any other philosopher of science's account, however, because discussion of the nature of science will unavoidably be carried on at such a gross level, consisting of wide-ranging and largely and unavoidably unsubstantiated generalizations.

We do not think that Kuhn's use of the notion of 'paradigm' is really meant to set one out on the meticulous classification of the different kinds and degrees of agreement that there might be within science, across and within disciplines. It is not a sociological term of art. The notion of paradigms is, in the first instance, meant to *highlight* a very stark contrast, between early stages in the development and later ones, and between the natural sciences (pretty much) and the social 'sciences' (pretty much): in other words, between the pursuits that have some kind of unity, as opposed to those that are in disarray. The 'exemplar' usage of the term highlights the degree to which there is *quite specific* agreement across the discipline or sub-discipline; the extent to which one piece of work can guide good practice, and enable the close relations that there can be between one bit of scientific inquiry and another in the same part of the discipline; the way in which studies in the natural sciences can often fit together in ways that studies in the social sciences seldom, if ever, do. (Without exemplars, no (real) science.) At the same time, this should not be overdone: the extent of agreement and disagreement is not to be treated as some absolute. As is plain, agreement is commonly and unproblematically (outside the world of philosophical fantasy) more or less: if one is involved in a relatively superficial transaction with others, then one might be in full agreement with them, but if one goes more fully into the terms of one's agreement one may find that the agreement is not so close as it seemed, or that there is much in the attempt at further and fuller specification of the agreement to disagree about.

Working on paradigms
(On section III of *SSR*, 'The nature of normal science')

The importance of paradigms (exemplars), in the initial instance, is that *they give scientists (real) work to do*. The fact that they contribute impressive solutions to existing problems is what makes them deserving of scientists' attention. That they solve some key problems is crucial to their attractiveness and acceptance, but another important source of their appeal is that they provide a rich source of *new* problems. Thus the paradigm is, we might say, a challenge – the challenge is to make it work as well as it can. Consider, for example, the attainments of the quantum physicists in the early part of the twentieth century, or even those of Darwin in the mid/ late nineteenth century. These contributions have provided problems that have kept large numbers of scientists seriously and purposefully occupied full-time at least into the early part of the twenty-first century, though paradigms do not normally last forever, and there assuredly will be further (conceptual) change. And this is the main element of Kuhn's historical reconstruction, discussion of the reasons why and the ways that paradigms displace one another. In *SSR* Kuhn emphasizes the way one paradigm would displace another within the same scientific specialism, but later came to think that paradigm change often involves the spawning of a breakaway specialism, and that this was the more important focus.

Solving puzzles and displacing paradigms
(On section IV of *SSR*, 'Normal science as
puzzle-solving')

We now reach the crucial point at which Kuhn's concept of 'normal science' is laid out. In order to understand why this is crucial, it will help to anticipate Kuhn's 'complementary' account of 'scientific revolutions'.

The introduction of a new paradigm into a science-with-a-paradigm is characteristically at the expense of the established paradigm (though separation and thereby greater specialization is another possibility), and successful installation of the new paradigm is the outcome of controversy. This displacement of one paradigm by another, and the controversy usually associated with it, is what Kuhn calls a 'scientific' revolution.

Scientific revolutions can, when completed, often be described as *total*. (This remark simply glosses Kuhn's important remark that scientific revolutions are *irreversible* (cf. *SSR*, 166).) However, it is equally important to Kuhn to stress that the grounds that produce such a clear-cut outcome are not necessarily themselves all that clear-cut. The 'received view' encourages the idea that scientists switch their loyalties from one scientific idea to another because they have established unequivocally that the new idea is better than the old, that new work decisively refutes prior work. Kuhn does not deny that the switch of a discipline's loyalties from an older to a newer idea, approach, etc., will *eventually* turn out to be absolute, and *looking back* it might, therefore, seem obvious to suppose that it was because of the plain, indisputable advantages of the new paradigm that it was universally preferred. But any such impression may well be entirely false to the historical record, and the choice between the disputed paradigms may have been anything but starkly obvious during the revolution.

Kuhn's *first* interest in 'scientific revolutions' then is in showing that the decisive results of these controversies may well stem from what were, *at the time*, less than conclusive reasons: the triumphs were not the 'knock-out' ones they might later seem. Any fair-minded comparison of the scientific rivals might give something rather closer to an 'on points' verdict, and recognize that the decision may have been 'a damn close-run thing' and akin, even, to a split decision. The fact that the verdict (for instance, in boxing) may involve a 'split decision' does not, however, make it any less final – its beneficiary is unquestionably the winner. Thus, it does not have to be – and in fact never is – that there is *nothing whatsoever* to be said for the scientific paradigm that loses out (see *SSR*, 99–100 and 107 for Kuhn's partial defence of the 'much maligned phlogiston theory' in chemistry). The victorious paradigm may well have won out over other contender(s) on *only a few points*.[7] The successful one may be neither 'completely successful with a single problem or notably successful with any large number' (23).

The differences – the *decisive* ones – between the paradigm installed as the new exemplar for up-to-date practice and the one it outmodes may be few and marginal ones from a point of view outside the science, but this is difference enough in the science. Kuhn's interest is in assessing what considerations played a part in driving the change *at the time*; his business is the depiction of the bases on which scientists satisfied themselves that they were doing the right thing.

Kuhn is not saying that one paradigm is demonstrably just as good as another, and denying there is any sense in which the election of one over another may ever be vindicated. It is worth remembering that Kuhn's concern is with reconstructing the historical situation *at the time*, without recourse to how things *later* turned out (his rejection of 'Whiggism').

Kuhn distinguishes between the fantasy of a completely successful paradigm and those that are in actuality encountered – relatively (more) successful ones (23). He echoes his more general conviction that the thorough exploration of nature can be pursued virtually indefinitely, and that continuous carrying through of the exploration will pose more – and more heterogeneous – problems than can be solved within any single framework of inquiry. Sciences attempt to capture the complexity of nature within a simple scheme, and the complexity of nature will always, in the end, overflow that scheme.[8] A new paradigm can be admirable or notable in that it can solve problems that are known to be more difficult than have been encountered before, or which have long proved intractable. It may identify a whole range of interesting new problems and have every prospect of satisfying them, but in all probability it will eventually encounter numerous problems that are not satisfactorily soluble in its terms.

Victorious paradigms, therefore, offer largely a *promise* of success, to which the achievement represents an initial guide. 'Few people who are not actually practitioners of a mature science realize how much mop-up work of this sort a paradigm leaves to be done or quite how fascinating such work can prove in execution' (24).

Normal science

The history of science after the formulation of a paradigm can be very roughly but profitably seen as an alternation of 'normal' and 'revolutionary' science. Normal science is that science which takes place on the basis of a paradigm, within a disciplinary matrix, and on the basis of accepted exemplars – the science done when the fundamentals stand beyond question. The idea of 'normal science' is one that can easily seem unappealing, making scientific work sound routine, dull and unimaginative, but this is a false impression. 'Normal science' is the condition under which most of the achievements of science are made, and the one under which the much vaunted accumulation of scientific results take place. Normal science is the expression of a humble truth at the heart of Kuhn's image of science – that investigation of nature is a complex task, most effectively pursued through a division of labour. It is only when the areas under investigation are 'typically minuscule' and where individual scientists operate with 'drastically restricted vision' – their attention entirely on their specific research studies,

not distracted from this by arguing over the fundamentals – that a
detailed and focused investigation of nature 'in ways that would
otherwise be unimaginable' (24) becomes possible.

Kuhn insists that during 'normal science' scientists are not in
search of fundamental innovations. They are working within pretty
well defined limits with respect to what can be brought into question – what *they need* to question. Their scientific activity really
amounts to the realization of the potential that the paradigm is
expected to provide, and the promise that drew scientists to the
paradigm to begin with. The development of the paradigm is *improvement on* its initial formulation, enhancing its precision and
extending its scope. This is where the cumulation in knowledge
takes place in something like the fashion envisaged by many
philosophers of science (especially prominently perhaps in Logical
Empiricism and its heirs), as a continuous addition.

To say that the aim of all this scientific work under conditions of
'normal science' is improving the precision and scope of the paradigm does not perhaps make perspicuous why scientists should
display 'the enthusiasm and devotion' (36) that they clearly have.
But Kuhn asserts that the individual scientist is *almost never*
involved in doing the things that people perhaps stereotypically
imagine is the greater part of scientific work, namely,

- opening up wholly new territory to investigation; or
- testing well-established belief.

(On this, see for instance, *SSR*, 37–8, 64–6, 77, 97 – and again
'compare' with sociology.) Scientists, according to Kuhn, are
normally preoccupied with the technicalities of solving the problems left over by an earlier, and very striking, achievement in their
area of work. What scientists find in their work is a challenge to
their ingenuity. Weinberg's charge, cited in our Introduction above,
that Kuhn has no explanation as to why scientists bother with these
problems is falsified. We might paraphrase Kuhn as saying that
people take up scientific problems because they find them deeply
intriguing, and badly want to investigate what is going on; in many
cases, they just can't leave these problems alone. Kuhn does not,
however, identify what Weinberg perhaps seeks, any further
purpose above and beyond the satisfaction of solving a difficult
problem; and this satisfaction is having solved a difficult problem
and thereby having made a contribution to human knowledge.

Before you react against the idea of normal science, pause – think
about those libraries full of natural scientific periodicals, and of what

the content of those must be. They must mostly fall somewhere between direct repetition of paradigm achievements and fundamental novelties. They can't be full of exactly the same stuff being done over and over. Sociologists of scientific knowledge have made a big, rather empty fuss about the fact that natural scientists don't replicate (much, if at all). Ask yourself who, with any minimal idea of how the natural sciences work, thought they did, imagined they were endlessly redoing each other's experiments? A stereotypical familiarity with the peer review process in a discipline like physics tells you that there are no rewards for coming second, that doing something over is of value only where something of importance hinges on it. A paper will get rejected just because the work has already been done: in Yazmina Reza's play *Life Times Three* (2001) terror strikes an astronomer when he is told that a paper on the very subject he is currently writing on has already been submitted to a journal.[9] His reaction: two years of his work has been rendered worthless.

Therefore, what is in the scientific periodicals must be stuff that produces novelty – it can't be straightforward repetition. At the same time, it can't all be ground-breaking, all-changing novelty: the kinds of 'fundamental novelties' we more or less non-scientific punters (who only keep up with the popularized stuff on science) do hear about are relatively few. Hence, most of what must be in those journals must be 'normal science': it does something that makes it worth publishing for the others in the same field, but it doesn't by any means turn everything upside down.

Thus, normal science is what Kuhn calls puzzle-solving (because it is – under normal science conditions – like ordinary puzzle-solving situations, where one is confident that there is, *that there has to be*, a solution and the only problem is to work out what it is) and the interest is only in those problems which can be assumed to have a solution. Problems will be set aside by scientists if it seems that they cannot be solved (37).

The analogy with puzzle-solving (36) is made to drive home the point that there are strong constraints on what it takes to solve a scientific problem in normal science. There are the conceptual, theoretical, instrumental and methodological commitments already in place within the scientific community (40). Also, scientific achievement demands a continual 'raising of the game': any result acknowledged to be an achievement must improve the scope and/or precision of the paradigm. The analogy with puzzle-solving is very important vis-à-vis saying what science is *for scientists*.

In short, Kuhn suggests that it is worth trying to see normal science *as* puzzle-solving – and that the results of doing so are illuminating, and have unfortunately been occluded from, and by, nearly all pre-Kuhnian (and much 'post-Kuhnian') philosophy of science.

On training and rules
(On section V of *SSR*, 'The priority of paradigms')

Kuhn now turns to the importance of *training* within the framework of normal science. The notion of the paradigm as 'exemplar' plays a part in challenging the idea that there is any 'scientific method' which could be specified as a set of rules prescribing in significant detail how a scientist should go about inquiry. No such set of rules is to be found spelled out in the scientific literature. Nor is instruction in such rules any part of training newcomers to a science, and they will then practise the science without having been taught any rules. Their training mainly involves confronting them with 'exemplars' (in the textbook, the lecture and the laboratory – or equivalent). It is through close study of these exemplars that trainees learn how to carry out scientific work (within their speciality).

There is one sense in which Kuhn is saying that science is dogmatic rather than critical, and here Popper voices a strong objection. However, we must be careful not to take the idea that it is authoritarian very far (cf. p. 113 below). Graduate training in the natural sciences can be dogmatic in the sense that students are presented with current science in a take it or leave it form: if they can't master and accept the current ways of doing things then they will not be admitted to a professional career in the field.[10]

In sum: paradigms are 'logically prior' to the research work that goes on within them. And, to continue with Kuhn's metaphor from Gestalt psychology, they are *the ground* against which innovations, anomalies, etc., can emerge as the *figure*. One sees anomalies 'against the background provided by the paradigm' (65).

Anomaly
(Section VI of *SSR*, 'Anomaly and the emergence of scientific discoveries')

According to Kuhn, the role of fundamental discovery, of fundamental factual or theoretical novelty, has been overstated.[11] Truly novel discoveries are not what are actually sought in the work of normal science, and the necessity for them is often recognized reluctantly. This does not say that normal science is hack work, that scientists are going for easy solutions – the work that is done in normal science is creative, productive and innovative, but innovation in the *fundamentals* of the discipline is relatively rare.

Fundamental novelty brings about changes in the way in which the science 'looks at the world': 'Assimilating a new sort of fact involves a more than additive adjustment of theory and until that adjustment is completed – until the scientist has learned to see nature in a different way – the new fact is not a scientific fact at all' (53). The change involved is, Kuhn's entire approach insists, neither a matter of accumulating, nor (*a fortiori*) one of accumulating *facts*. The change is of a kind that Kuhn sometimes terms a change in 'worldview'. In part this change alters the considerations that delimit what could possibly be accepted as a fact within the discipline. The change is not one that involves new findings as such, but one which – in accord with the idea that it involves a paradigm shift – involves a change in the idea of what properly scientific problems are and how they may be solved.

The appearance of anomaly

The key to fundamental novelty is, for Kuhn, the occurrence of 'anomalies', a term that precisely captures the implication that novelties are novelties *only relative* to some paradigm, are things which *do not fit* the existing scheme.

An example is what occurred in the 1770s vis-à-vis chemistry. Here is Kuhn: 'In 1774 [Priestley] identified the gas [produced by heated red oxide of mercury] . . . as common air with less than its usual quantity of phlogiston. . . . Early in 1775 Lavoisier reported that the gas obtained by heating the red oxide of mercury was "air itself entire without alteration [except that] . . . it comes out more pure, more respirable"' (53–4).

One can see the burgeoning anomaly right there, in Lavoisier's peculiar, almost tortured language.[12] What Lavoisier eventually proposed – and what, as it happens, Priestley could never accept – was that the gas being produced here was not something which could be neatly fitted into the boxes provided by the paradigm of the time, 'phlogistic' chemistry. Kuhn concludes that 'Only when all the relevant conceptual categories are prepared in advance' can we intelligibly speak of discovery as a point event (55). The revolution from 'phlogistic' to 'modern' chemistry overturned conceptual categories, and so, Kuhn suggests, it is misleading to depict it as happening at one particular place and time, or being carried out by a single person. (One could describe the discovery of (say) xenon like that, once the periodic table had become well established – but not the discovery of oxygen, for which no place had been prepared in the chemistry which preceded it.)

No change without something to change to
(On sections VII and VIII of *SSR*, 'Crisis and the emergence of scientific theories' and 'The response to crisis')

Here Kuhn takes a first step towards the expression of what seems to many a very troubling idea – the rejection of the idea of a 'fixed nature' (see p. 58 below).

A crucial point in Kuhn's argument, one that is broadly 'Pragmatist' in nature, is the idea that 'radical critique' alone is an idle wheel in science (77). Scientists only give up an accepted paradigm when there is some alternative they can attach themselves to: that there are some things the paradigm cannot do does not detract from the fact that there are many things it can do. That there are things which the paradigm cannot do is, Kuhn is suggesting, a normal situation, even a necessary situation in something that is (still) a science, and not, *of itself*, a fateful flaw. The fact that there are things that do not fit an existing paradigm does not result in withdrawal of the paradigm, which makes it plain why there are anomalies – if scientists followed the strict 'logic of science' (*à la* Popper for example, a logic of refutation, where a single negative instance can – ideally – invalidate a whole theory), then, when they found something which did not fit with the paradigm they would reject the paradigm and go back to square one, meaning, of course, that there would be no such things as anomalies in Kuhn's sense. But this is not what scientists do.

Kuhn on the chemical revolution example again: Many things lose weight upon being burned. Well, at least, they *appear* to. If one investigates very carefully, collecting all the ash and the water vapour released and the smoke particles etc., one finds that they become slightly *heavier*. As Kuhn writes, '[Lavoisier] was much concerned to explain the gain in weight that most bodies experience when burned or roasted, and that again is a problem with a long prehistory. At least a few Islamic chemists had known that some metals gain weight when roasted' (71).

Ah, so there was a clear refutation of the phlogiston theory available, and it had been available for ages? Not so fast. While it is true that 'In the seventeenth century several investigators had concluded from this fact that a roasted metal takes up some ingredient from the atmosphere', still 'that conclusion seemed unnecessary to most chemists' (71). Why? Well, 'If chemical reactions could alter the volume, color, texture of the ingredients, why should they not alter weight as well? Weight was not always taken to be the measure of quantity of matter. Besides weight-gain on roasting remained an isolated phenomenon. Most natural bodies (e.g. wood) lose weight on roasting as the phlogiston theory [said] they should' (71). Long-standing anomalies are usually just – things to ignore.

When weighing became more accurate (leading to more and more cases of weight gain), and when 'the gradual assimilation of Newton's gravitational theory led chemists to insist that gain in weight must mean gain in quantity of matter', then phlogistic chemistry started to look bad. *Even then*, phlogiston was not done for, 'for that theory could be adjusted in many ways. Perhaps phlogiston had negative weight, or perhaps fire particles or something else entered the roasted body as phlogiston left it' (71). But phlogistic chemistry became less and less *attractive*, especially to newcomers to the discipline.

It requires more than the mere existence of anomalies to set scientists to re-examining the fundamentals, searching for solutions outside what the paradigm allows.

So, what do scientists do when their discipline seems to be in some kind of crisis?:

> [We should note first] what scientists *never do* when confronted by even severe and prolonged anomalies. . . . [T]hey do not renounce the paradigm that has led them into crisis. They do not, that is, treat anomalies as counter-instances, though in the vocabulary of [Carnapian, Popperian, etc.] philosophy of science, *that is what they are*. (77; emphasis is added)

We hope it is now obvious that this does not mean that theories in science can proceed merrily along, without *regard* for how things are in the world, for how one's experiments are going, etc. There is no reason why anyone should misread Kuhn's claim that the history of science has never yet revealed anything which resembles 'that methodological stereotype of falsification by direct comparison with nature' (77) as suggesting that comparison with nature *has nothing whatever to do with it*. Comparison with nature takes place *in the context* of the comparison of paradigms, and on the terms provided by these. In a period of normal science, there is regular comparison of the paradigm's expectations with nature, for this is, of course, what puzzle-solving often consists of: seeing how well the paradigm works out in further cases. What else are anomalies except the cases in which the paradigm's expectations *are un*satisfied, in which nature *does not* behave according to these expectations, and the scientists can understand that this is not what they were expecting?

The search for fundamental novelty is provoked, if at all, by anomalies, but, since anomalies always exist without necessarily provoking such quests, the question remains: what makes an anomaly worth concentrated scrutiny? It depends on the specifics of the case. There is no algorithm for fundamental scientific change: this is a point which fundamentally disappoints rival philosophers like the Logical Empiricists or Imre Lakatos. In the 'extraordinary' period in science leading up to a scientific revolution, some scientists no longer depend on and work within the paradigm in the same unquestioning way, but, in their attempts to work out just what it is about the anomaly that is anomalous, attempt to sharpen the tension between the anomaly and the paradigm. They are thus apt to put the usual practices deliberately under strain. Since the capacity of the paradigm to serve as a reliable guide in exploring the area of the anomaly is what is in doubt, the scientists' behaviour will be less well directed than under 'normal science' conditions, and will be a bit more like the random casting about characteristic of the pre-paradigmatic case (with even occasionally explicit argument over fundamentals, or behaviour more like that of a philosopher than of a normal scientist (87–9)).

What are 'scientific revolutions'?
(On sections IX and X of *SSR*, 'The nature and
necessity of scientific revolutions' and 'Revolutions as
changes of worldview')

In an attempt to clarify what is involved in the substitution of one paradigm for another Kuhn makes an ultimately somewhat ill-fated analogy with the 'Gestalt switch' in which people are able to alternate between two discrete perceptions of the same thing. The 'Gestalt switch' is commonly identified in psychology by the 'duck/rabbit' in which a schematic drawing can alternately be seen as a duck and a rabbit, or by a picture in which the image of two faces alternates with that of a vase.

It was an analogy Kuhn cautiously made (albeit hardly cautiously enough). The analogy's value is in emphasizing that paradigms are rivals in the sense that scientists can accept either the prevailing paradigm or its proposed alternate as valid, but cannot simultaneously entertain or accept both (85). The central limitation on the analogy is that in the usual case people can switch back and forth between the two perceptions – now they see the duck, then a rabbit, and then revert to perception of it as a duck again. Scientists cannot engage in such reversion. When they move from one alternate to the other they have given up the first for the second. The switch is strictly one way, there is no going back. It is a permanent one-time-only Gestalt switch. Another disanology applies – the Gestalt idea involves talk of people 'seeing things as' this or that (for instance, now as a duck, now as a rabbit), which has an inappropriately provisional character in comparison with the categorical ways in which scientists express themselves: they do not, at least when committed to a paradigm, speak of themselves as seeing things 'as this' or 'as that', but just assert that they see those things (85, 114–15).

At last, scientific revolutions

Scientific revolutions are those times[13] at which one paradigm replaces another, or, as Kuhn later came to emphasize, a new area of research spins off from an established one on the basis of a new exemplar. Obviously, consideration of how these revolutions take place is critical to Kuhn's attack on the received image. It ought now to be clear that Kuhn will certainly decline to accept that a scientific revolution is a dispute between an obviously right party on one side and an obviously mistaken one on the other. Since the paradigm provides the means for settling disagreements in scientific results, if the paradigm itself is in dispute, then the usual means – the *only* means – for settling disagreements are out of order. During such revolutionary periods, the situation in the science may be more like the pre-paradigmatic condition than it ever is in periods of normal science: fundamentals are in question, there are meaningful possibilities of fundamental novelty, there is a lack of focus and a sense of casting about within the community. However, while this is more *like* the pre-paradigmatic situation, this is not a *return* to any such condition, and is certainly not going to involve any starting completely afresh and all over again.

The founding analogy – with political revolutions – is seriously and multifariously intended, but should not be taken too far (and Kuhn's rhetoric perhaps gets a little strong on *SSR*, 93, for example). The analogy's main value to Kuhn is to provide a reminder that the conflict between the defenders of a political status quo and their revolutionary opponents is one that cannot be resolved by neutral, authoritative adjudication. Where the society once had authorities that would settle disputes, in a time of revolution there is no longer any authority that is recognized by both sides in the struggle, and the only resolution possible is therefore the outright defeat of one or other party.

One of the key things that Kuhn wants to say about scientific revolutions is that their dynamics have to be understood at the level of the scientific grouping, rather than as a matter of individual choice (again like political revolutions, they involve choices between 'incompatible modes of community life' (94)). Considered at the level of the scientific community, scientific revolutions don't take place through – certainly not *only through* – switching of allegiances on the part of individual scientists. In Kuhn's considered view (this is obviously largely an empirical question), many individual scientists just don't *switch* allegiances at all. Those who have been trained and pursued their careers in the established paradigm may not give up their allegiance, and their resistance to attempts at change makes the attempt to bring in the new paradigm a revolutionary *struggle*. Equally, the protagonists of innovation have not switched their allegiances either – most *never were* attached to the older paradigm, but have entered the profession with the proposed innovation. Thus, the revolutionaries are often made up of younger scientists.

This is not to say that individuals *can't* switch. Kuhn's argument is that understanding what such individuals do is not the exclusive key to understanding scientific revolutions. Such revolutions are shifts in the collective balance within an area of scientific work, where a small minority may enter and eventually triumph over what was previously the as-near-complete-consensus-as-you-will-ever-get. Here, the idea is that the revolution takes place within the same field of scientific work, that the old paradigm is thrown out and the area of work reconstituted on the basis of the new paradigm.

We don't have to worry too much about whether *all*, or even large proportions of scientific revolutions are really like Kuhn's picture. (Nor indeed worry about which candidate revolutions were 'really revolutions'. Kuhn is providing one with a tool-kit, not an encyclopedia of truths – cf. pp. 91–2 on *BB*, below.) The relevance here is that the transition between one paradigm and another is not necessarily a matter of what has traditionally been imagined to comprise a rational choice. To the extent that it involves a changing balance in the composition of the profession, it is not a matter of *individual* choice at all. Neither, of course, is it a collective *decision* in the sense that all have been parties to a collective agreement: it is simply an emergent outcome of the exigencies of the struggle between the rival camps. Kuhn is mocking the idea of the community arriving at a decision by means of explicit *rational and conclusive debate*. This is not to say that there is not debate, for there surely is, still less to deny that there is rationality, of which 'science' *even at times of revolution* could arguably be seen as a set of paradigms; but the idea of scientific debate effectively and purely 'rationally' persuading people from the old to the new view is, at best, an over-simplification. There is plenty of debate, but where the debate affects people, and induces change, it does not do so in the way that the received view imagined. The change in those who do switch allegiance is (notoriously, as Weinberg complained) more like a religious conversion than a rational, that is impartial, reflection on and appraisal of two rival points of view. Kuhn's question is how far this debate features actual and direct disagreement rather than *arguing in circles, begging the question,* or *just talking past each other.* This is why scientists can't simply, by good logical and evidential proofs, establish to each other's general satisfaction which of two rival positions is correct.

The debate in scientific revolution, as in political revolution, is often circular. There are real obstacles to mutual persuasion since each party is appealing to principles that the other contests. For example, in order to accept a conclusion you have to subscribe to the premises it follows from, but this is just what the parties involved do not do (consider the face-off between the divine right of kings and the principle of democracy in France around 1789–92). In the scientific case, each group depends on its own presuppositions to justify and evaluate its results, but a logically convincing proof can be given only to someone who concedes the premises to begin with, and disputants cannot therefore *prove* to their rivals' satisfaction that they – the rivals – are wrong. This explains why there are many who are unaffected by the arguments of their rivals. Argument can provide a clear and vivid display of the vision of the new paradigm, of what scientific practice will be like for those who adopt it, and some people may respond to this. It cannot, however, be made compelling 'for those who refuse to step into the circle' (94). So, it is hard for controversialists in these debates *to understand each other*, to appreciate each other's point of view, and, Kuhn argues, the fact is that in one way or another, they *often don't*.

While the scientists are adamantly refusing to change sides, they may also be under misconceptions about what the other position actually is. It is not that the controversialists find each other's positions hard to believe, but they often find them hard to understand in that they cannot see that what the other says makes any real *sense*. This is a crucial innovation of Kuhn's: emphasizing that deep scientific disputes involve questions of sense/meaning just as much as (in fact, more than) questions of true and false, questions of fact.[14] If this is so, then the two sides – to some greater or lesser extent – are handicapped in explaining their respective positions to each other, and in appreciating each other's points of view.

Does it *have* to be this way? Must it really be that some new phenomenon or theory absolutely cannot be assimilated to the existing paradigm? If the anomaly could be smoothly integrated into the existing paradigm, then the stereotype of science that Kuhn has rejected, as developing in a 'fully cumulative manner', would be true. But the stereotype does not fit the facts: 'cumulative acquisition of unanticipated novelties proves to be an almost non-existent exception to the rule of scientific development' (96). There are good reasons for this. Normal science research *is* cumulative, but novelty of the sort at issue here can only exist to the extent that it does not square with the logical consequences of the paradigm in place. So, there *must* be a conflict (a 'logical gap'; a gap, that is, between two more or less logically coherent but distinct systems) between the existing paradigm in terms of which the anomaly is truly that, and the new paradigm in relation to which the phenomenon is no longer an anomaly but, rather, among its logical derivatives (97). The differences between them, further, are not just substantive. The new paradigm brings about reorganization (or 'cannibalization') of old science in the relevant discipline, perhaps reallocating some of its problems to another discipline, declaring others unscientific, and promoting things previously deemed not to be problems, or only trivial ones, to pride of place (103). In the Copernican context we will remark on Kuhn's talk of the 'hard core of knowledge' that remained constant across scientific changes, noting the extent to which the content of the previous paradigm is never entirely abandoned, with 'old science' carried over into the new context. However, in line with the Gestalt-switch analogy, there is in *SSR* much greater, or at least more explicit (than in *The Copernican Revolution (CR)*), emphasis on the extent to which the 'preserved old science' changes its character: what is carried over will be extensively altered by transplantation.

Once again we need to bear in mind Kuhn's emphasis on the issue of the complexity of paradigm-to-paradigm comparison. It is not that there are no quite general criteria which may be used for invidiously comparing paradigms,[15] but we can list four considerations:

(a)   in real science people are characteristically already signed
      up to one or another paradigm when they make these
      comparisons;
(b)   there are real obstacles to properly identifying the char-
      acteristics of the respective paradigms in the revolutionary
      situation;
(c)   there is an *indefinite plurality* of general criteria, with any one
      paradigm scoring well on only *some* of these;
(d)   in any case, there is no formula which dictates which of
      these criteria should be given priority or how they should
      be traded off. Using the same criteria people can end up
      drawing quite different conclusions as to which is – in an
      overall judgement – the better paradigm. There *are* rules for
      comparing paradigms, but no formula for applying the
      rules conjointly.

II
World changes
(SSR, 111–35)

In the latter sections of *SSR*, Kuhn starts to say some strange-
sounding things, which seem to many at least troubling, if not outright bizarre. Having considered how paradigms 'constitute'
science – that is, give order and structure to its inquiries – he
announces that he now wants to explore the 'sense in which they
[paradigms] are constitutive of nature as well' as science (110).
Rather than hastening to agree with or dissent from this claim,
readers might pause to reflect that it is already *qualified* in the light
of being said to apply 'in a sense'. Rather than flatly and forcefully
affirming that 'paradigms are constitutive of nature', Kuhn is *himself*
pondering what, in saying this, he might mean – and does say that
he is not yet sure what he means: 'I am convinced that we must
learn to make sense of statements that at least resemble these'
(121).[16] Next, another crucial, but also crucially qualified statement:

> Examining the record of past research *from the vantage of contemporary
> historiography, the historian of science may be tempted to exclaim* that
> when paradigms change, the world itself changes with them. Led by
> a new paradigm, scientists adopt new instruments and look in new
> places. Even more important, during revolutions scientists see new
> and different things when looking with familiar instruments in

places they have looked before. It is rather as if the professional community had been suddenly transported to another planet where familiar objects are seen in a different light and are joined by unfamiliar ones as well. Of course, nothing of quite that sort does occur: There is no geographical transplantation; outside the laboratory everyday affairs usually continue as before. Nevertheless, paradigm changes do cause scientists to see the world of their research-engagement differently. In so far as their only recourse to that world is through what they see and do, we *may* want to say that after a revolution scientists are responding to a different world. (111, emphasis added)

It is clear, noting that Kuhn is speaking of what the *historian* might conclude, and if one attends carefully to his phrasing, that Kuhn is largely but not entirely aware that saying that 'when paradigms change, the world itself changes with them' is a manner of speaking only. He is uncertain as to whether this only restates what he has argued thus far, or says something rather more than this. It has got some quasi-metaphysical anxieties attached to it – what we will shortly identify as the phenomenal worlds problem.

Kuhn's formulation is acceptable *only in so far as* he, Kuhn, is searching for a form of words that rids us of our temptation to misunderstand the history of science – he is not issuing a metaphysical thesis. What he gives here is a formulation that emphasizes (perhaps even exaggerates for thought-provocative purposes) the impact scientific revolutions have on the outlooks of scientists, to help in overcoming the received view.

For the scientist, the world might be said to change. As a *descriptive* characterization of the differences which the historian, comparing two successive periods, notices, it is unproblematically intelligible: it is *as though* there had been a change in nature itself, so very different are the pre-scientific and post-scientific environments. Before the revolution the scientists would talk prominently about a certain phenomenon that after the revolution they would never mention again. And, reciprocally, after the revolution, scientists start talking about phenomena that had never figured in their discourse before. However, though these are only strong metaphors to give a flavour of the changes paradigm shifts engender, Kuhn is often understood as implying more than this, as if he were saying in a literal way that 'paradigms do constitute nature', and that, therefore, when a paradigm changes *the world literally changes with it*. Thus, one can see how the 'world changes' issue gains its name.

We are not saying, at this point, that Kuhn *is wrongly understood* as intending something more literal than metaphoric with talk of such world changes, for we think he *is* drawn to the idea that more than a *façon de parler* is involved. Even so, we think that his work is clearly dissociated from any such drastic conception as that changing a scientific theory brings about changes in the natural world itself. However, there is an important change of gear taking place here, as the notion of 'different worlds' slips from being an idiomatic expression into a philosophical one, where 'world' often means much the same as 'reality' or, even, 'natural reality' – as in 'there is a world out there.'

Enter phenomenal worlds

Kuhn's attraction to saying that the world changes with paradigms leads him into a problematic as a result of the way he conceptualizes perception, and, therefore, observation. Kuhn is driven towards that conceptualization because of the implications of his arguments about paradigm shifts, considered in relation to the empiricist tradition in philosophy of science from which he is trying – but not quite managing – to make a decisive break.

Given Kuhn's views about the historical relation between an earlier and a later paradigm, he feels compelled to say that, descriptively, the Aristotelian perception ('of a pendulum's motion') is just as accurate (119) as Galileo's – and, indeed, that the triumph of the latter over the former was something of a swindle. Galileo's preconceptions about pendulum motion 'led him to see far more regularity than we can now discover there' (119). From Kuhn's point of view, it seems we have to say that *both* an Aristotelian and Galileo himself, observing the same case of motion (a stone swinging on a string), will make observations that are, in their respective terms, *largely, if not entirely accurate.* Given Kuhn's anti-Whiggism we cannot resort to the otherwise ready solution: to take Galileo's description of the case as identifying what is there to be observed, and deeming, therefore, that the Aristotelian has failed to observe correctly to just the extent that his perception deviates from Galileo. So, we can't say that one of the two parties, Galileo, correctly observed the facts, and the Aristotelian failed to do so. Without some independent and definitive source which says what the facts are, how are we to say whether one person has observed the facts more correctly, more accurately, than the other? We can actually say, or Kuhn does, that the two describe the facts equally well. Thus, we seem compelled to say that each party did observe what they reported themselves as observing – but each observed something different in the same place.

Having made an attempt to deal with the issue by allowing that each did observe what they said they observed, Kuhn has intensified, not eased, the pressure on his situation. What are we now to say about two scientists, confronted with the same phenomenon – a swinging stone on a string – observing something different: cases of constrained fall and pendulum motion respectively? Once we have got this deep into a hole, the Wittgensteinian philosopher would advise that we stop digging, but Kuhn is not that much of a Wittgensteinian. He keeps digging. Kuhn seeks his way out of his difficulties by declining to treat the question 'do they observe the same thing or something different?' as calling for a yes *or* no answer, and proposes instead to answer it with: yes *and* no, they do and do not see the same thing.

This alone is OK, but Kuhn thinks he now needs an account of perception/observation. He adopts what we will call a 'two moments doctrine'; in a philosophical terminology Kuhn does not himself use, this involves the staple of empiricist thought, the 'given' and its interpretation. Kuhn wants to escape from the untenable consequences of belief in a given – this is what his terminological contortions are about – but often, as here, he falls back into thinking that *some kind* of world, or some set of sensations, *is* given. There *is* only one physical world out there, and Kuhn (with virtually everyone else) wants to say that *this* does not change when scientific theories do, and he never suggested otherwise. *Therefore* there is one and the same natural world, having its natural effects on the scientists observing it, and, since they are biologically similar creatures, affecting them in much the same way. *There is* one and the same shining light up there in the sky that scientists from different astronomical paradigms *do see*. So each sees something, and in these terms, they both see *the same thing*. However, we cannot say in a scientifically neutral way what each sees, just because the identity of the (source of) light is what they are disputing. What each can possibly say they see in scientific terms is not at all the same thing, and whether one says a planet is observed and the other that it is a star will depend entirely on their respective scientific traditions. *Without the appropriate tradition*, one cannot say that one sees a planet or a star, and what, therefore, the scientist 'actually observes' is *a composite*, made up of the input from nature, and the input from the paradigm.

Kuhn therefore feels driven to say this: 'Until that scholastic paradigm was invented there were no pendulums but only swinging stones for the scientist to see. Pendulums were brought into existence by something very like a paradigm-induced gestalt switch' (120); but is puzzled enough by his own expressions to have to ask, 'Do we, however, really need to describe what separates Galileo from Aristotle . . . as a transformation of vision? Did these men really *see* different things when *looking* at the same sorts of objects. Is there any legitimate sense in which we can say that they pursued their research in different worlds?' (120).

Kuhn feels forced to continue with these strange locutions, though 'acutely aware of the difficulties I am creating' in so doing. He sees no choice but to reaffirm that 'though the world does not change with a change of paradigm the scientist afterward works in a different world. . . . I am convinced that we must learn to make sense of statements that at least resemble these' (121).

Still he does not convey a sense of a clear and confident understanding of what he is using these strange locutions to say. Kuhn is perhaps clearer on what he is trying to get away from than of what he is trying to put in its place, but maybe he has not yet got far enough away from some of the things that philosophers of science used to say, and as a result is creating a puzzle for himself, imposing strange expressions on himself. Perhaps this is substantially a good thing. Kuhn's locutions are, he thinks, forced on him by his desire to avoid the language of the largely discredited empiricist tradition, and (more generally) of the 'received view' in the philosophy of science. Kuhn is trying to *avoid* the doctrine of 'the given'. This is something that he has *in common* with most major recent philosophers, notably Sellars and Davidson (though Davidson misses this agreement, and so misses Kuhn). Our point is that: Kuhn has not *found* a way of avoiding the myth of 'the given' that does not yield new philosophical perplexities.

The empiricist tradition had not depended on the 'hypothetical fixed nature' but rather 'fixed experience' as the 'court of final appeal' in its analysis of scientific dispute. The notion of an 'observation language' distinct from the 'theoretical language' of a science is meant in philosophy of science to be what describes scientists' experience, describing what they observe in a way which is 'neutral' between competing theories.[17] Rival theories can be compared with experience described in the observation language, a language that is independent of both their terminologies. But, following Kuhn's arguments, this cannot be true – we will see, in Copernicus' case, how deeply (in Kuhn's view) the 'interpretation' goes relative to the observation. Indeed, given the 'doctrine' of the 'two moments of perception', the interpretation penetrates the observation. (If the two moments are only notionally separable, this is another way in which Kuhn genuinely *endeavours* to free us of the myth of the given.) Kuhn is arguing that if one is going to talk about anything that might be called scientific observation one cannot effectively disentangle the given from its interpretation. What counts as an observation in science is, for Kuhn, a composite of a given input from nature and of the interpretive capacity supplied by training in a scientific paradigm.

Furthermore, the 'given' in the laboratory should actually be termed 'the collected with difficulty' (126). It takes a full and deep absorption of the paradigm, plus the acuity and intelligence required to practise normal science effectively, before a scientist is in any position to make observations that are of interest to other scientists on the frontiers of research. Near those frontiers especially, there is no way that scientists' experiences can be decomposed into any more elemental form – direct reports of bare sense experience, divested of all the paradigm-based learning that is the precondition of the scientist being let loose in the lab at all (126–8) – that allows specification of what they observed in a way which does not draw on the 'theoretical' language of their paradigm. Kuhn insists that scientists are right to treat such things as pendulums and oxygen as 'fundamental ingredients' of their immediate experience, and not as theoretical construals of any more basic experience. This means though that there is no stable or common experience that can be appealed to in adjudication between paradigms either, because the experience itself is partially composed of the interpretation. Here Kuhn is most definitely not saying that one *first* perceives an input from nature, and *then* gives it an interpretation. What he is saying is: the scientist's perception is composed of both the input from nature and the interpretation. For the purposes of scientific observation, then, the experience cannot possibly be extricated from the interpretation, *and therefore* the scientists cannot have recourse to the pure, extricated element of 'the given' in their experience. Thus, the observational materials – the data – available to scientists cannot be said to remain stable across changes of paradigms, because the observational materials are *irreducibly* characterized in terms which draw on a paradigm. Thus: 'The data themselves had changed. [W]e may [therefore] want to say that after a revolution scientists work in a different world' (135).

Kuhn denies that the staple of the empiricist approach, a distinction between observation language and theory language, can be of use in understanding scientific change, but he has not thereby entirely abandoned the distinction, between the given and the interpreted, that the empiricist's distinction was meant to express. He has not succeeded in abandoning the myth of the given. He has retained in some measure the distinction between the given and the interpreted, but shifted the point at which the dividing line between them can be drawn. The line is no longer drawn between what is given in raw perception, providing 'sense data' and the interpretation that is then applied to that 'given', for, on Kuhn's account, the given and the interpreted are combined in perception, in what can meaningfully be called – in respect of *scientific observation* – *sense data*. Our view is that this is progress; though better still to find a way to *give up* the whole idea of 'the given'. However, it is retained, and its retention provides encouragement to maintain that two disputing scientists both do make genuine observations, for while they are observing something different they are nonetheless observing something that is the same: the same bits of the natural world enter into their perceptions in each case, and it cannot then be said that one scientist does observe something, and that the other does not observe anything at all. Both observe something that is real, but the way they experience what they observe differs (because of their different backgrounds). Therefore, *neither* observes the external world of nature *in itself*, but it cannot, on those grounds, be said that they make invalid observations. At the same time, the different experiences that their observations generate cannot be invidiously contrasted, for there is no way of saying that one observes *nature in itself* more closely or accurately than the other. Therefore, there is the temptation, to which Kuhn succumbs, to say that the respective scientists each observe something real, that the world observed by one is no more or less real than the other, and that, therefore, each observes a reality, each lives in a real, but different, world. The worlds found in the respective scientists' experiences are not, however, to be confused with the real world *in itself*: one way in which to express this (in accord with the Kantian connection) is to say that the scientists inhabit *phenomenal realities*, that they occupy different *phenomenal* worlds.

What the beginning scientist learns to see is determined *jointly* by the environment and the particular normal-scientific tradition that the student has been trained to pursue (112, 113). If *part* of that perception is dependent on changes, namely, the contribution of previous 'visual conceptual experience' because important – especially conceptual – parts have been withdrawn and replaced, then the scientist's perception must be re-educated, which among other things means acquiring a new Gestalt. This is another reason why discussion between different paradigms is always *at least slightly* at cross-purposes.

These considerations, whatever the merits of particular forms of words via which Kuhn (and we) try to adduce them, are clearly meant to fill out the analogy with revolution, and with the absence of any authority to adjudicate between the two sides. Kuhn has eliminated the empiricist's only possible 'final authority' – raw sense data – and, given the general pre-eminence of empiricism in Anglo-American philosophy of science, has virtually deprived philosophy of science of the notion of any authority external to the scientific schemes themselves.

The idea of a 'fixed nature' can play no part in what Kuhn envisages as a historical understanding of scientific change. The idea of 'world changes' is not, itself, a denial that there is any such thing as a 'fixed nature', compatible with a superficial interpretation of the claim that when the world according to science changes, nature itself changes with it. 'The world out there' retains a certain constancy throughout: stars do not transmute into planets, mixtures do not mutate into compounds. However, while one may be confident that nature does not alter with each paradigm shift, there is no way of saying *what it is that remains constant* throughout. The question of what it is that remains constant throughout is the very question over which paradigms are contesting each other: is it, and was it always, a planet or is it really a star; is it, and has it always been, a compound, not a mixture? Thus, while it might be accepted that there is a 'fixed nature' in this sense, it is not one that can be appealed to as an independent point of reference against which competing paradigms can be compared, that can play the kind of role as a final authority that philosophy of science requires of it. To try to say, in any substantive way, what is actually 'out there' as the object of scientific contest would be to make an (at least tacit) identification with one side of the argument or the other. A 'fixed nature' is a useless ornament on, not a working addition to, Kuhn's model of scientific

change. He is saying that the idea of a 'fixed nature' is only an ide-
alization and plays a distorting role in understanding these problems – that is, the problems of the philosopher and the historian,
not necessarily the problems of the scientist themselves. His appre-
ciation that he is resorting to a 'strange locution' indicates Kuhn's
unease.[18]

The effects of incommensurability
(On sections XI and XII of *SSR*, 'The invisibility of
revolutions' and 'The resolution of revolutions')

Incommensurability – which literally means 'cannot be compared
by a common measure' – has been involved in the argument about
the discontinuities between successive paradigms in the discussion
both of paradigm shifts and of the 'world changes' issue. There is
no independent, neutral standard against which two competing
paradigms can respectively be compared. They can only be compared with each other, though the idea of doing that is complicated,
and does not provide the kind of comparison that philosophers of
science need – one that adjudicates which of the two is most closely
in accord with empirical reality. The idea that philosophers can
make that kind of comparison, standing transcendentally outside
the hurly-burly of scientific dispute, is being exploded.

Perhaps the most important of *all* Kuhn's thoughts with respect
to philosophy of science is on whether economics is the most successful social science because economists know more about science
and truth than, say, sociologists do. Or is it that they know more
about economic matters than sociologists do about society? Gener-
alized, this question asks: could philosophers, who know about
'truth' and about 'science', possibly know more about what scientists should do than the scientists, who know, after all, nothing
about 'truth' and 'science' in the sense that philosophers do, for the
former only know about black holes, or quarks for example. From
Kuhn's point of view this is a rhetorical question.[19]

We have not yet paid much attention to the *second* respect in
which incommensurability matters to Kuhn, and that is as a source
of misunderstanding between competing scientists. Far from rival
scientists taking the true measure of each other's positions, Kuhn
maintains, they often misunderstand each other's views, and, as a
result, in scientific revolutions sometimes do not really critically
engage with each other, just talk past each other – and after the

revolution the losers are simply forgotten: 'communication across the revolutionary divide is inevitably partial,' he remarks (in)famously, at one point (149). To see just what Kuhn means here, however, let us see what else Kuhn actually says. He remarks:

> at times of revolution, when the normal-scientific tradition changes, the scientist's perceptions of his environment must be re-educated – in *some* familiar situation he must learn to see a new gestalt. After he has done so, the world of his research will seem, *here and there*, incommensurable with the one he had inhabited before. That is another reason why schools guided by different paradigms are always *slightly* at cross-purposes. (112; emphasis added)

And now a lengthy quotation, because the issue of incommensurability is so important. First:

> all historically significant theories have agreed with the facts, but only more or less. There is no more precise answer to whether or how well an individual theory fits the facts. But questions much like that can be asked when theories are taken collectively or even in pairs. It makes a great deal of sense to ask which of two competing theories fits the facts *better*. Though neither Priestley's nor Lavoisier's theory, for example, agreed precisely with existing observations, few contemporaries hesitated more than a decade in concluding that Lavoisier's theory provided a better fit of the two. (147; emphasis in original)

And, shortly after:

> If there were but one set of scientific problems, one world within which to work on them, and one set of standards for their solution, paradigm competition might be settled more or less routinely by some number of problems solved by each. But, in fact, these conditions are never met completely. The proponents of competing paradigms are always *at least slightly* at cross-purposes. Neither side will grant all the non-empirical assumptions that the other needs in order to make its case. Like Proust and Berthollet arguing about the composition of chemical compounds, they are bound partly to talk through each other . . . we have already seen several reasons why the proponents of competing paradigms fail to make *complete* contact with each other's viewpoints. Collectively these reasons have been described as the incommensurability of the pre- and post-revolutionary normal-scientific traditions. . . . In the first place, the proponents of competing paradigms will often disagree about the list of problems that any candidate for the paradigm must solve. Their

standards of their definitions of science are not the same. Must a theory of motion explain the cause of the attractive forces between particles of matter or may it simply note the existence of such forces? Newton's dynamics was widely rejected because, unlike both Aristotle's and Descartes's theories, it implied the latter answer to the question. When Newton's theory had been accepted, a question was therefore banished from science . . . more is involved, however, than the incommensurability of standards. Since new paradigms are born from old ones, they ordinarily incorporate much of the vocabulary and apparatus, both conceptual and manipulative, that the traditional paradigm had employed. But they seldom employ these borrowed elements *in quite the traditional way*. Within the new paradigm, old terms, concepts, and experiments, fall into new relationships one with the other. The inevitable result is what we must call, *though the term is not quite right*, a misunderstanding between the two competing schools. . . . to make the transition [from Newton's] to Einstein's universe, the whole conceptual web whose strands are space, time, matter, force, and so on, had to be shifted and laid down again on nature whole. (147–8; emphasis added)

And, finally,

the third and most fundamental aspect of the incommensurability of competing paradigms [is that there is] a sense that I am unable to explicate further, [in which] the proponents of competing paradigms practice their trades in different worlds. One contains constrained bodies that fall slowly, the other pendulums that repeat their motions again and again. In one, solutions are compounds, in the other mixtures. One is embedded in a flat, the other in a curve, matrix of space. Practicing in two different worlds, the two groups of scientists see different things when they look from the same point in the same direction. Again, that is not to say that they can see anything they please. Both are looking at the world, and what they look at *has not changed*. But *in some areas* they see different things and they see them in different relations one to the other. That is why a law that cannot even be demonstrated to one group of scientists may *occasionally* seem intuitively obvious to another. Equally, it is why, *before they can hope to communicate fully*, one group or the other must experience the conversion that we have been calling a paradigm shift. (158; emphasis added)

The reason for these extensive quotes is that they show that Kuhn's *SSR* formulations of the idea of incommensurability do not themselves settle a key issue, namely, just how great are the misunderstandings between scientists? Is it the case that scientists from different paradigms are hardly able to make sense of each other *at all*, that they misunderstand each other on almost every point, no matter how small or great? One could read these passages as licensing such a reading, yielding a picture of scientific revolutions as almost farcical episodes of mutual incomprehension. *On the other hand*, one can read the same passages as claiming something much more modest, and less far-reaching, namely, that incommensurability means that failure of mutual understanding is an exigency of revolutions, that there almost certainly will be times when minds do not meet, but that these will be occasional, on particular points, either major or minor. Kuhn eventually clarified that the latter reading was his intended and preferred one.

From these remarks, one thing is however certain: Kuhn is not saying that incommensurable theories *cannot be compared* – what they can't be is compared in terms of a system of common measure. He very plainly says that they can be compared, and he reiterates this repeatedly in later work, in an effort (mostly in vain) to avert the sometimes catastrophic misinterpretations he suffered from mainstream philosophers and postmodern relativists alike. Integral to – though tacit in – his point in saying this is that paradigms are characteristically *complex constructions* and that comparison of them is a multidimensional affair; he is considering theories as ('conceptual') schemes rather than, for instance, unconfirmed conjectures. Thus, it is not as if a theory can be individually accepted or rejected simply on the basis of fitting the facts, because, as we have explained before, in Kuhn's view any scientific scheme that has had any support both will and will not 'fit the facts' (79–83). There are always abundant – but largely inconsequential – anomalies to any theory, but there are also *lots* of instances that – in its terms – fit the theory too. Hence, the question cannot be asked, *does this theory fit the facts*, the only meaningful question is: does this theory fit the facts *better than this other one*. So, the two can be compared, and it can be *eventually* fairly conclusively decided by the science – as in the case of Priestley versus Lavoisier – that the latter's approach fitted the facts better than the former's. This offers no comfort to Whig historians or traditional philosophers of science. To say that Lavoisier fitted the facts better is *of course* what Lavoisier's heirs (that is, all of us, in so far as we have chemical knowledge) will say.

The importance of incommensurability for Kuhn is not that it presents a problem for science. It does not. The misunderstandings between scientists are virtually never mutually experienced as such;

it is not that scientists will be brought to a grinding halt by the realization that they really cannot understand each other, and that they need to establish reciprocal understanding before they go further. It is important in other connections for Kuhn that the scientists do not usually realize that they are talking at cross-purposes. They think that they are in straightforward disagreement, that they understand the other's position well enough, and that they can see just what is wrong with it. Thus, it may be on the basis of (some) misconceptions about the usurped paradigm that its replacement wins out, but the verdict remains irreversible.

Incommensurability presents a problem only for the philosophy of science. Kuhn is denying – as a cumulative result of the arguments about revolutions, world changes and phenomenal worlds – that it is possible to compare scientific theories in the way that philosophers of science imagined they could, that they can be matched – by those with a only a spectator's interest in science, such as philosophers – against each other in some decisive way. Kuhn is denying that two theories can be jointly compared against any independent, neutral standard or set of facts, and also – and this is the distinctiveness of the notion of incommensurability – that the two theories can possibly be lined up against each other as though disagreeing on the answers that they give to a list of common questions, allowing us to assess which gives the right and which gives the wrong answer to each question.

Even to speak of Kuhn 'denying' claims in the philosophy of science may be misleading. For Kuhn aimed usually to establish the *nonsensicality* of the claims he was rejecting. In roughly the following sense: he aimed to get his interlocutors to see that there was nothing that they really could want to mean by a point-by-point comparison of theories across major changes in science; that they would fail to understand the predecessor science as science, and/or simply fail to understand it, if they insisted upon such comparison, or 'translation' (discussed in chapter 5; see also 'Commensurability, comparability, communicability', in *RSS*, for Kuhn's attempts to persuade his opponents (in this case, Kitcher, and, by extension, Davidson and Quine) that they should give up their anti-incommensurabilism when they realize that it makes a complete nonsense of any attempt to render phlogistic chemistry in terms of our own, and still be understanding phlogistic chemistry).

If a point-by-point comparison is imagined as a matter of placing the assertions made by the empirical part of rival theories in 'conjunction' *with each other* and with cogent evidence bearing on the truth of each, then point-by-point comparison would involve (for instance) comparing Newton's answer to the question of what was the cause of attraction between particles with Descartes's or Einstein's answers to these questions (see *SSR*, 148). Aristotle says *this* is the cause of the attraction, Newton says *that* and Einstein says *the other* – now which, if any, is the nearest to being right?

However, Newton does not give *any* such cause but not because he overlooks doing so. Newton does not accept that his predecessors have identified the cause, but not because they identified the wrong cause and he is now about to identify the right one.[20] He disagrees with them over whether you *need* to answer the question, whether your account is significantly incomplete if you don't feature a cause of attraction. And that of course brings up in turn other differences between Newton and the rest. There is comparison here, but it is not 'point for point'. For the difference between Newton's theory and its predecessors – and its successor(s) – is not that each has a different answer to the same question, but that Newton's scheme *leaves no space* for the question that the others ask.

We mentioned the need to be wary of the use of the term 'coincide' to represent the point at which there are empirical results that can be compared between paradigms, even when paradigms look as though they are in respects directly comparable in a point-to-point way. This may be a false impression. It is one of the key suppositions of Kuhn's whole enterprise that resemblances between paradigms are often superficial only – the paradigms only *seem* to be in agreement, and the scientists adhering to them are making assertions that only look as if they are mutually empirically contradictory, and thus straightforwardly comparable (if one assertion contradicts another, then they are at least in the same ball park).

Take the case of Kuhn's remark (102) that mass *à la* Newton and *à la* Einstein can be measured in the same way at low velocity but 'must not be conceived to be the same'. This is surely illustrative of what Kuhn means by incommensurability. It might be objected against us (and Kuhn): If the measurements given, i.e. the quantities specified, are specified in the same units before and after the Einsteinian revolution and (in those units) give the same values, then how could they not be the same? Certainly Kuhn will not deny that these units and measures appear identical, but he is arguing that this is just the kind of confusion that motivates his whole enterprise – they may look as if they are just the same, but one had better not treat them as if they actually are. The difference is not in the values yielded, but in what it is that the values yielded *are measuring*. It is perfectly correct, from an Einsteinian point of view, to say

that the correspondence in quantifications with Newtonian measurements is good enough *at low velocities*. However, this is itself an
Einsteinian, not a Newtonian way of talking. To say that Newtonian calculations are correct when measuring mass at low velocities
is, in Newtonian terms, to misrepresent what Newtonians were
doing. Mass is a fixed quantity in the Newtonian scheme, and so
when it has been correctly measured or calculated, that is what mass
is. The velocities at which mass was measured were not conceived
in the Newtonian scheme as 'low' velocities, for there was no physically relevant conception of high velocities – such as approaching
the speed of light – with which to contrast them.

So, though it is perfectly correct in Einsteinian terms to say that
the Newtonian scheme works perfectly well for masses measured
at low velocities, this is not a correct characterization of what, in
Newtonian terms, was going on. When Newtonians use 'mass', its
'grammar' is of a fixed quantity which is conserved. Just think of
the difference between 'football' in 'association football' and 'rugby
football', and between 'goal' in the former, and 'goal' in the latter
(and in 'American football', too). Going 'over the bar' in one qualifies it as a goal, in the other, it disqualifies it as one. Wrestling with
'reference' does not really help, we think; the difference is in what
you can say and mean in the language and in the way in which
expressions fit in with other expressions: *in the context* of rugby football, 'It's over the bar' means 'He's scored', in that of association
football, it means 'He's missed'.

The idea of Newton and Einstein being rivals is a *restricted* one
if it is taken to mean that one must make contradictory claims,
when, in fact, the rivalry is of a different kind – the Einsteinian calls
for reformation of the Newtonian vocabulary so as to raise questions that couldn't be asked within Newton's scheme, and thereby
displace Newtonian usage, so that for instance Newtonian 'mass'
becomes 'mass at low speed' (which is certainly not what it originally meant). Thus, it is OK to say that 'the two theories remain conflicting accounts of the same thing'[21] so long as we remember that
this means they are conflicting accounts of mass, and that mass in
Einstein is by no means simply the same thing it is in Newton! There
is no contradiction in saying this – if one understands Kuhn's 'position' appropriately.

In sum: Newton and Einstein are definitely in *competition* with
one another – that's why you can *only* have one of them, not one as
a special case of the other. But, to help avoid confusion we must
insist that Newton and Einstein do not *contradict* each other – *not*

because they agree (at all), but because the nature of their 'disagreement' involves them necessarily talking across each other (see 149 and 98).

So, what is plain about Kuhn's presentation of 'incommensurability' is that it raises issues that have to do with the mutual comprehensibility of rival paradigms to the scientists who are involved, in a partisan way, with them.[22] Clearly, incommensurability has considerable bearing with respect to the actual, not the *perceived* understanding of some scientists by others. If an Einsteinian scientist employs Newtonian calculations to determine mass-at-low-velocities, this is not actually using the Newtonian methods, since those do not calculate mass-at-low-velocities, but just calculate mass. *If* the scientist is indeed operating in this way, then she or he is working on a misconception, and is using not the pure Newtonian ideas, but only those ideas reconstrued in Einsteinian terms. Of course, it makes no material difference to the scientist's science (only to their image of their activity, their sense of their position in history, etc.) that this way of calculating is used as a convenience, or that in order to do this it is stripped of its specifically Newtonian aspects. It is not, either, at all possible for the Einsteinian to do otherwise. Einsteinians know that it makes a difference whether mass is being accelerated through low velocities or at very high ones, and they cannot competently operate 'the Newtonian techniques' save on the basis of this knowledge. There cannot be the sense, for Einsteinians, of calculations of mass, period, but only of calculations of mass-at-low-or-high-velocities. As used by Einsteinians, these calculations are *in the historical sense* not-really-Newtonian, only 'partially' Newtonian. Thus, the claim that Einsteinian physics has incorporated Newtonian physics as a special case is erroneous, for it has rejigged Newtonian physics or, in another metaphor, cannibalized it; or, as we ourselves would put it,[23] has *reconfigured the grammar*.[24] Einsteinian physics is a different language-game: though of course 'language game' is a term involving the *activities* with which these words are bound up.[25] Science isn't just words.[26]

Scientific development
(On section XIII of *SSR*, 'Progress through revolutions')

We now come to the last of the strange or outrageous things that Kuhn says in *SSR*, and one of the main topics which has been provocative of most subsequent debate, namely, the sense in which science can be regarded overall as 'making progress'. In a way, Kuhn's previous discussion has straightforwardly and effectively answered the question as to the sense in which science can be said to make progress – his description of the way in which paradigms replace one another (in concert with his earlier description of the accretion of knowledge in science under normal conditions) itself *describes the way in which science can be said to make progress.* What Kuhn has already done is, in a way, to describe how the scientific specialists determine that they are making progress by showing how they opt for one paradigm over another. Once this has been understood, further questions about progress – and progress 'towards Truth' – are rendered empty.[27] By excluding 'truth' – as it has traditionally functioned in the philosophy of science – and making, rather, such observations as those listed, Kuhn is effectively rejecting the idea that science is the fulfilment of the aims of metaphysics. Realists are apt to treat science as though it has delivered or will deliver what metaphysics sought for, namely a depiction of what there ultimately or finally really is, and their question then is how close is science to rendering this final, ultimate description? Kuhn doubts that this metaphysical notion is needed in characterizing the results of science, meaning that he cannot pose, let alone answer, the question that the Realists would want him to.

Now, Kuhn holds that the assertion that 'science makes progress' is to a considerable extent a circular one, for something does not qualify as a true science unless it makes progress. But how can such a claim be justified? If we are to accede even to this claim, do we not want some independent measure of progress to ensure that 'making progress' is different from 'just substituting one paradigm for another'? But this is just what we cannot have. If we understand the relationship between scientific work and the scientific community then 'the phrases "scientific progress" and even "scientific objectivity" may come to seem in part redundant' (162). One can't – at least Kuhn says he can't, and doesn't believe he needs to – find a way, independently of that scientific community, of saying whether a science is making progress. It is that 'independently of that scientific community' which is the crucial component. It is not as if scientists decide to replace one paradigm with another and then ask themselves whether, having done so, they have made progress. A paradigm is being displaced because the relevant scientists have decided that some alternative to it *is progressive*, and have decided this on the basis of the considerations integral to their science itself. Thus, viewed from within any scientific community, with its set of problems, strict standards, checks and balances, the result of its work *just is* progress.[28] This is not quite a matter of tautology, but of understanding the extent to which, for any science, *the content* of any notion of progress is provided by the practice of the science itself, and by the way that criteria for progress just are those which count for and against candidate paradigms. A mature science possesses 'unparalleled insulation' from external standards because the application of meaningful standards of assessment depends upon a familiarity with the technical complexities of the science. There is an intense collective assessment of work but this is done by the *cognoscenti*. The science decides both what the worthwhile problems are, and what can count as a solution to them. How can someone who cannot understand just what, scientifically, is going on (such as a 'methodologist of science' who is not actually a practising scientist in the field) contribute to this?

Further, saying that the scientific community judges progress overlooks the fact that the scientific community itself is defined through the replacement of paradigms, through the (eventual) marginalization and elimination of those remaining unregenerately attached to a sidelined paradigm. The scientific community is made up of the (current) victors, and they control the history of the science, automatically regarding their triumph as the measure of progress. This is not, however, to offer any cynical view of the matter as though might simply makes right, for to say that would be to ignore the fact that these are *scientific* revolutions, and that dispute over the *scientific* merits of the rival paradigms is the stuff of their opposition, that the triumph consists in the reconfiguring of the science itself. So, we do have scientific revolutions, and Kuhn *can* be called 'the philosopher of scientific revolutions', although he could equally well, perhaps even more so, be called 'the philosopher of ordinary science'. But this gives no support to those in sociology who think that now Sociology (or Philosophy) can lord it over the sciences: and it gives no support to those in the Science Wars who would like to *overthrow* the epistemological authority of science. Kuhn was right to be deeply wary of his 'followers'.

We might gesture at how progress in science can be indicated just in the expansion of the sciences, their division of labour, the dense content of their textbooks, and the sophisticated level of their puzzle-solving. Such gesturing is gross and scientifically entirely superficial, and therefore quite different from being able to say, in any *particular* instance of paradigm substitution (in any situation where the question as to where to stand in the science is still 'live'), what the specific features of the transition are which merit the description of the substitution as one of progress, unless one is in possession of considerable competence in the technicalities of the science in which the displacement is taking place.[29] To engage substantively in the comparison of the two paradigms would be *to engage in the scientific argumentation itself*, not in some separate philosophical adjudication of scientific argumentation.

So Kuhn is more 'moderate' than he has been seen by friends and foes in philosophy and the social sciences; but he is more 'radical' in a wholly different direction. Radical in the way that Wittgenstein, Peter Winch, much ethnomethodology (and much Foucault) are radical.

2

The Historical Case Studies

When we were planning this book, we did not intend to devote much space to the exposition of Kuhn's case studies, but we were forced to change our mind. Kuhn's historical case studies are, we insist, very important to understanding the meaning of his work overall, and to getting a clear picture of what *SSR* actually means. We would urge those interested in knowing what Kuhn has to say about the philosophy etc. of science *to read (some of) Kuhn's historical studies*, if they are able to do so, and to use this chapter as an 'orientating device'/companion to such readings. The two (book-length) studies we principally discuss in this chapter – a chapter which we hope will establish securely our sense of the salience of Kuhn's historical work to his work on philosophy of science – are the main ones; and are both revisionist histories. Kuhn claims to have shown that Nicholas Copernicus and Max Planck each failed to solve the technical scientific problem they had set themselves, and from which resulted the scientific work that gave them the reputation for bringing about scientific revolutions. Kuhn further argues that neither Copernicus nor Planck actually produced the scientific achievement that earned them their revolutionary title: Copernicus did not establish that the earth moves around the sun, and Planck did not actually introduce the modern idea of the quantum into twentieth-century physics. Kuhn thus deals with two major periods in the formation of modern science, the so-called Copernican revolution perhaps marking the beginnings of modern science and certainly of modern astronomy, the quantum revolution marking the break between classical and modern physics. If Kuhn can justify his claim that two such momentous instances of scientific change have been extensively misunderstood, this is surely proof positive for his general thesis about the risk of misunderstanding in the historiography of scientific developments generally.

Kuhn's treatment of the key figures, Copernicus and Planck, is meant to do justice to each of them by achieving a faithful understanding of their actual contributions to science, and respecting the integrity of their scientific thought. Doing justice, though, involves divesting them of the credit for their supposedly most momentous achievements, to the extent that Copernicus and Planck (and possibly several other scientific heroes besides) were in central respects barely revolutionaries at all.

The Copernican revolution

Kuhn argues that the revolution in astronomy replacing the earth-centred conception of the cosmos was only in *(comparatively) small part* due to anything original to Copernicus. Copernicus' (vital) contribution was *to give mathematical cogency* to the (by no means unprecedented) idea that the earth itself might be mobile, as an accidental by-product of his attempt to *preserve*, not undermine, most of the previous Ptolemaic scheme of astronomical thought. The Copernican revolution belonged as much to those coming after Copernicus, and they were not necessarily enamoured of Copernicus' efforts: hence the actual revolution in astronomy took place over quite a long time, and involved other important contributors who were not necessarily motivated by the desire to promote Copernicus' work.

Kuhn's *The Copernican Revolution (CR)* devotes relatively little attention (some 50 of 280 pages) to Copernicus. Over half the book is given over to persuading us of the *plausibility* of pre-Copernican astronomy to those who lived with it. That conception, essentially a 'two-sphere' one, combined Ptolemy's astronomy and Aristotle's physics, and it was such that, Kuhn seeks to show, 'the Ancients' were every bit as justified *by their lights* in thinking themselves the possessors of the truth about the universe as we now think ourselves as being.

- The Ptolemaic system was, observationally, very successful and highly technical astronomy, with a good quantitative grasp on the positions and movements of the stars.

- The theoretical structure of the Ptolemaic system provided a coherent and plausible interpretation of the observational record.
- The idea that the earth could be a planet was pretty near *meaningless* within the Ptolemaic tradition.

By the fourth century BC there was a dominant consensus that the earth was a sphere suspended at the centre of a much greater rotating sphere which was the location of the stars. The sun was within the space between the inner and outer spheres. This was the 'two-sphere' model which, though not unique, became the predominant conception. (Just to avoid possible confusion, let us note that there are two different ways of using the expression 'two spheres'. One is to talk of the earth being the innermost sphere, and the spherical movement of the stars the outermost, with Venus, the Sun, Jupiter, etc., being ('on' spheres) in between. A different use, more associated with specifically Aristotelian ideas, is that the earth and the heavens are drastically different realms in respect of the forces which govern them[1] – the sublunar realm, including the earth, has a drastically different nature than the superlunar realm, and these two realms (or spheres) are governed by different principles.)

Kuhn argues that this conception translates well into naked-eye, looking-up-into-the-sky observation, which was all that was available to observers before Galileo and the telescope, and insists, for example, that the idea of a dome-shaped sky is compatible with observation to the extent that, looking up into the sky, it does indeed 'look domed'. If it is further conceived that the semi-spherical appearance of the sky we can see is only half of a completed sphere, then there is intellectual closure to the idea. In addition, if – as it was – the idea of symmetry is itself greatly valued, then the idea of the sphere also offers a perfectly symmetrical shape, and this could be deemed a sign of spiritual perfection.

Thus, Kuhn asserts (*CR*, 30): 'In astronomy, the two-sphere cosmology (in the first of the above two senses) works and works very well.' He means by this that the two-sphere model systematically captures a multiplicity of the diverse geometrical relations among astronomical observations, as he seeks to show by working through some details of the scheme. *That it is a scheme* is, for Kuhn, an important fact, recognition of which encourages acceptance of a crucial general point about the nature of science (which will bear – see below, p. 148 – on the critical issue of 'incommensurability').

Conceptual schemes, on Kuhn's view, provide compact sum-
maries of observational materials, interconnecting those that would
otherwise be unrelated, and thus easing the burden on memory
in that specific observational facts need not be remembered, but
can be derived from the scheme. The conceptual scheme goes
beyond the facts it summarizes, for from it can also be derived conclusions about how (if one accepts the premises of the scheme)
various other things ought to be, making it 'a tool for predicting
and exploring the unknown' (CR, 39). The two-sphere universe
was sound enough as a summary of observation to be used, and
an economical enough device still to be used *even today*, as the basis
for navigational computations, in preference to the more complex, less economical (with respect to computational effort) post-
Copernican scheme. *At the time* the two-sphere universe was also a
fertile source of predictions and inspiration to research, but then scientific inquirers could (indeed, virtually *had* to) accept its premises.
That, of course, is no longer the case. *At the time*, however, the
two-sphere model both fitted very well with the observations of the
naked eye, and seemed like a perfectly natural way to explain
things (that is, it fitted with both the observations and also with
other then generally influential ideas (religious, philosophical,
scientific)). The idea that the earth was in motion would seem
then absurd, almost inconceivable.

Having argued for the plausible way the two-sphere scheme
fitted the framework of thought at the time, Kuhn then notes the
worm in the bud: that the two-sphere scheme, best condensed in
Ptolemy's astronomy, '*never* quite *worked [completely]*'. Ptolemy had
brought together an apparatus of ad hoc devices into

> the first systematic mathematical treatise to give a *complete, detailed,*
> and *quantitative* account of all the celestial motions. Its results were
> so good and its methods so powerful that after Ptolemy's death the
> problem of the planets took a new form. To increase the accuracy or
> simplicity of planetary theory Ptolemy's successors added epicycles
> to epicycles and eccentrics to eccentrics, exploiting all the immense
> versatility of the Ptolemaic technique. (CR, 73; emphasis in the
> original)

Given the two-sphere scheme, and the importance attached to the
idea of sphericity that reflected the value of symmetry, it was
natural enough to resolve the problem of planetary motion by distributing the planets among concentric orbits in between the inner
and outer spheres. *Given* that the outer sphere of the universe was

a rotating shell, it was easy enough to infer that the planets were also contained in similar shells, and this is where the physical structure of the universe is relevant. Simple circular motion was insufficient to capture the irregularities of observed planetary motion, and so the astronomical system acquired a further apparatus of deferents, epicycles, erratics and equants, that 'elaborated' on circular motion.

A deferent is a circular orbit centred on the earth, which a planet is following as it moves around the earth. However, it can be conceived that the planet is also involved in an additional circular movement, relative to the deferent, this circle being centred on the point at which the planet is located on the deferent. This circle is an 'epicycle'. The introduction of epicycles means that the planet traverses an orbit even as it follows the general line of the deferent, and its overall pattern of movement is not simply circular, but acquires a looping structure whose precise character will depend upon the number of epicyclic orbits the planet is conceived to complete during a complete rotation around the deferent. This basis for calculation could be used to eliminate the grossest anomaly of observed planetary motion – retrogression (going back on its orbital motion) – but other, lesser anomalies remained. 'Erratics' and 'equants' were both used to reconcile the general model with observed behaviour. Erratics were deferents that were not centred on the earth, which would, of course, give a distinctive geographic configuration to the pattern of planetary motion, while equants were contrived to accommodate the apparent variations in the rate of planetary motion.

The equant involved a different assumption from the one that held that the motion of the planet should be constant relative to the geometric centre provided by the centre of the earth. Instead, the rate of motion was assumed to be constant relative to a point displaced from the earth. It was this apparatus of ad hoc devices that Ptolemy brought into a relatively coherent scheme, but even the apparatus of erratics, equants and the rest could not regularize *all* the observational data. These recalcitrant observations – which, when sufficiently worrisome, constituted what Kuhn later termed 'anomalies' – were not taken as a basis for throwing out Ptolemy's whole scheme, but rather as a reason for persisting in it: they gave astronomers an apparently worthwhile task. Persistent efforts to refine the system, by Ptolemy and many able successors, could not engender a version of the system that 'ever withstood the test of additional refined observation' (*CR*, 73). In the words of *SSR*, the

Ptolemaic system was a notable success, but never a complete one. In those of *CR*: it never quite worked.

In *CR*, this spawned two complementary questions for Kuhn: Given the 'success' of the Ptolemaic scheme, why was it eventually overthrown? Given the persistent 'failure' of the Ptolemaic scheme, why was it not overthrown long before? These are the questions that arise for Kuhn from the situation as he has pictured it, but they are not ones that can be answered by the history of the scientific ideas themselves. They pertain to what he calls 'the anatomy of scientific belief' (*CR*, 72–6, 182f.), to the things that affect scientists' inclination to continue in or abandon scientific beliefs. The fact that many scientists accepted the Ptolemaic system is, presumably, accounted for by the fact that it was such an impressive achievement, *to which no one had proposed a comparably impressive alternative*, and one which was, in the absence of alternatives, *prima facie* persuasive. And once the basic suppositions have been accepted, then the anomalies seem to require only minor – albeit elusive – modifications to accommodate them. Apart from this general tendency to 'inertia', it was a feature of the Ptolemaic scheme that it was interwoven with the mightily prestigious physics of Aristotle.

The conceptions of motion and of the four elements in Aristotle's physics reinforced the idea that the earth must be at the centre of the universe. Within the universe there is no vacuum; all of space is occupied by matter. The substance of the universe outside the orbit of the moon is the aether, out of which is made the planets and stars, and the concentric shells within which they are distributed, and whose motions propel them. The celestial and earthly domains are different: the former is a domain of perfection and permanence, the latter a realm of change and decay. The celestial realm is formed from one pure, unchangeable substance, the aether, but the earth and its immediate lunar space is formed of four elements which (because of the moon's movements) are intermixed, rather than, as they would otherwise tend to, separating themselves out. Within this context, we find the heart of Aristotle's conception of motion, and of the natural tendency of elements and the bodies they make up to seek their natural position – if the four elements were not continually intermixed, then fire would have risen to the highest, outermost level, and earth been at the lowest. Thus the natural motion of earthly composites is towards the centre of the universe (a universe assumed to be spherical), and it is therefore in that position that the earth must reside. It would be against the nature of the element of earth to move from the centre.

Astrology, then immensely widespread and influential, also backed up Ptolemy's scheme: astronomers and astrologers were often one and the same, for astronomical observations were used to construe the significance of the heavens for life on earth. In short, there were, at the time, plenty of well-entrenched reasons, in science and in religion, to find the two-sphere worldview entirely credible, and though people did, even then, propose the idea of a moving earth, they could not offer a rationale for it that could disturb this mutually reinforcing complex of reasons backing the Ptolemaic scheme.

The passage of many centuries before the 'rediscovery' of the Ptolemaic tradition, after the long decline in learning between 2 BC and the twelfth century, loosened the interconnections between these mutually supporting elements.

The rediscoverers laboured under misconceptions about the historical depth and diversity of 'Ancient thought', and thus were apt to treat temporally remote and intellectually very different traditions as if they were elements in a single system, with divergences between them being looked upon as logical discrepancies (CR, 100–5). Thus, for example, the mutually supporting interconnection of Aristotelian and Ptolemaic thought was eroded in connection with astronomy, where the qualitative character of Aristotle's arguments was seen to conflict with the quantitative attainments of the Ptolemaic tradition. The rediscovery further led to the incorporation of the Aristotelian cosmology into the Christian one, though there were already critics of Aristotelian astronomy, whose work anticipated Copernicus – ones, however, who did not gain much recognition, nor achieve any great impact, until after Copernicus. This background, plus the intellectual turmoil created by the wide-ranging and multifarious social changes of the Renaissance period, provided a context in which there could be reduced resistance to fundamental rethinking because of the way the turbulence was loosening the hold of established ideas generally. It also created conditions that would ensure receptiveness to Copernicus. Indirect 'humanist' influences on some natural scientists (including Copernicus) inclined them towards a belief in the importance of mathematics in understanding nature, and a view of the sun as 'the source of all vital principles and forces in the universe' (cf. CR, 128–33). This had affinities with the rather mystical views of mathematics and about the sun as an expression of deity which were derived from Plato by some of Aristotelianism's philosophical opponents.

It was not internal developments within astronomy, then, that triggered the first stage of the Copernican revolution, but diverse historical and social changes.

> [T]he very process of rediscovery, the medieval integration of science and theology, the centuries of scholastic criticism, and the new currents of Renaissance life and thought, all had combined to change men's attitude toward the scientific heritage that they learned in school. Just how great, *and yet how strangely small*, this essential change could be we shall discover. (*CR*, 132; emphasis added)

Copernicus' own intervention, according to Kuhn, was not meant to overthrow the Ptolemaic system but essentially to restore (and improve) it. The very attempts of Ptolemy and his followers to make their scheme a complete success had attenuated the integrity of the whole, effectively tending to make it disintegrate into a plurality of different schemes, bringing about a scientific scandal. Copernicus wanted to reunify the Ptolemaic system by making *relatively* minor adjustments of a mostly *mathematical* character to it. The proposal that the earth be allowed to move, that the earth be reconceived as a planet, was not necessarily intended as anything other than a purely mathematical step to effect a closer fit with observational data on a more *mathematically* coherent basis. Neither Planck nor Copernicus intended the step that earned them their revolutionary reputation as a serious empirical hypothesis, only in the first instance as a device to achieve a technical solution, namely, in Copernicus' case, to get rid of the 'equant' which he regarded as one ad hoc device too many. The presumptions which had originally held the Ptolemaic system in place had, as we have said, been loosened by social and other change in Europe after its rediscovery, but Copernicus himself was not in a scientifically significant way affected by these changes, says Kuhn, because they were external to his very narrow and specific focus. That was on *precisely what it would take to get the mathematics right*, without thought for any further consequences or their perturbing effects.

Copernicus had, then, a quite specific problem within the Ptolemaic scheme, the problem of the planets, that is, getting the formulae of the system exhaustively to capture their apparent motion. This was, says Kuhn, both a highly technical and apparently petty problem, far removed from the everyday life that it was eventually to affect significantly. But if it was a petty problem, why was Copernicus making such an effort with it? Kuhn's answer: while it might be a small anomaly 'in itself', it could be an intense irritant to someone of a mathematically precise turn of mind. It made the astonomical system seem 'monstrous', and generally unsatisfactory.

The innovative aspect of the Copernican system over the Ptolemaic is dealt with in five pages, and these are not all that complimentary. In respect of cosmology, Copernicus' system adhered to the two-sphere universe (in the first sense only). And one could dispense with the major epicycles if 'retrograde' planetary motion was transformed into a merely apparent phenomenon resulting from the orbital motion of the earth.

Copernicus's supporting arguments for his proposal were not adventurous. They largely align him with the Aristotelians, from whom, in proposing that the earth be considered to move, he makes only limited deviations. Indeed, he sought to make out that the proposal of a moving earth was an application of the core principle of the tradition, that spherical shape is intrinsically associated with motion. If, then, the spherical shape of the earth is shared with the celestial bodies, why should not the earth also share in the ubiquity of circular motion? Nor were these arguments *prima facie* all that convincing: Kuhn virtually sides with those who made mock of Copernicus for the weakness of the previous supporting arguments, and further argues that Copernicus' model was only arguably more economical than Ptolemy's. Copernicus dispensed with major epicycles and with equants, but he retained epicycles and erratics, and required as many circles (over thirty) as Ptolemy to achieve results which were themselves often only approximations to observations.

The real advantage the Copernican system *initially* had was not an astronomically practical one, but – Kuhn holds – mainly aesthetic. The geometrically neat and coherent account yielded by Copernicus' sun-centred system would appeal to those who could recognize that neatness. This particular strength (most crucially in making the order of distribution of the planets and the relative size of their orbits calculable from observation, without the addition of further, rather arbitrary assumptions as required in Ptolemaic astronomy) appealed only to a few astronomers, but gave them the motivation to develop Copernicus' work into the simple and accurate solution that he *to some extent wrongly* claimed.

The place of the planets within the 'two-sphere' scheme was the critical 'moment' of Copernicus' intervention. We can see here what might later tempt Kuhn into forms of expression that proved

hostages to fortune, but can also see why these locutions might
seem useful. We could (and not falsely) say that Copernicus' astronomical achievement was to make the earth a planet. Also (again,
not falsely): prior to Copernicus, the earth was not a planet, but after
his work was accepted, it was. If the (deflationary) tone in which
such a suggestion is made is appreciated, it will be clear that Copernicus is not being credited with making a heavenly body go through
some kind of physical transformation, turning it from one kind of
thing into another. Copernicus is being credited with changing the
outlook of astronomy, altering the way that we identify something, fundamentally reclassifying the earth (in astronomical terms),
making *thinkable* what had previously been unthinkable (and, eventually, vice versa).

Thus, before Copernicus, the earth was not deemed a planet, it
did not satisfy the current criteria for classification as one. The
planets were identified not by their material composition, but by
their motion; 'planet' meant 'wanderer', and the movements of the
planets, relative to those of the stars, wandered. Since the earth was
the *fixed* centre of the universe it did not wander and so *could not be*
a planet. Some of the Ancients, including Aristotle, held (for good
reason) that the earth was a sphere, but not a moving one. The earth,
being at the centre of the two-sphere universe *could not* move. Thus,
in the pre-Copernican scheme, the earth was not a planet. The effect of
Copernicus' proposal to view the earth as moving would be to
change its fundamental classification; it changed from being (in the
Ancient scheme of things) a body that did not move, to one which,
like Saturn, Jupiter, Venus and others, did move. It *had 'become'* a
planet. But why use this form of expression at all, for it is surely
risky? Well, this form of expression provides a thumping emphasis
on Kuhn's point about the difference between the pre- and post-
Copernican ages: in the earlier age there was rock-solid confidence
that the earth was not and could not be a planet, while in our own
time there is equally solid confidence that it *is* a planet. On such
rocks is science built.

Eventual publication of Copernicus' ideas in the year of his death
did not precipitate any immediate and strong reaction. His work
was largely inaccessible to those without technical competence (see
CR, 185), and among those it could be recognized as an impressive
achievement, but that was not necessarily enough to encourage
adoption of its radical proposal as a serious hypothesis. Copernicus' work and techniques could be used without accepting that the
earth really did move, and over the next generation they were to

find increasing use among astronomers, further eroding resistance to the idea of a mobile earth.

Copernicus could argue that the mathematics required for his calculations were compatible with the view that the earth's orbit was at around or near the centre of the spherical universe, but this had a knock-on mathematical, and therefore, eventually, astronomical consequence; it required a considerable increase in the ratio of space between the outermost planet and the stars. The occurrence of a vast gap between the outer planet and the stars effectively deprives the nested shell version of the universe of (the neatness of) its rationale (but does not *refute* it), and opens up the opportunity to question the rational purpose of so much empty space. Copernicus provided a basis for rendering the two-sphere conception problematical, but it was others who raised questions against it that Copernicus himself never explored, and was not necessarily even aware of making possible.

Copernicus' successors (over about 150 years) brought out these implications and the downfall of the Ptolemaic system. Tycho Brahe opposed the idea of the earth's motion, but both his observational work and his investigations of transient phenomena such as comets further weakened the authority of the traditional cosmology of the unchanging heavens. These latter observations did not depend upon any astronomical breakthroughs, and could have been made long before. In the new context, they acquired a significance they did not have before. Kepler, a convinced Copernican (and, like Copernicus, of neo-Platonist persuasion), was critical of, and rationalized the scheme's structure. Seeing an inconsistency between the residue of traditional astronomy which retained a somewhat special place within the system for the earth, and the main achievements of the scheme, Kepler reworked the mathematics to give equivalent treatment to all planets, thus producing a simpler and more accurate scheme. Attempting to rise to the new, higher standard of observation set by Brahe, Kepler eventually found that the solution to many outstanding problems arose from replacing the circular motion (which Copernicus had held as dear or even dearer to his heart than his predecessors) with elliptical motion. At last the epicycles and erratics could be dispensed with. With Kepler, the Copernican revolution was nearly 'complete'; Kuhn finds its final completion in Newton's system.

Galileo's introduction of the telescope into astronomy provided (we can say, retrospectively) a final blow against the Ptolemaic tradition as an equal contender in the debate. Kuhn argues, though, that the observational evidence produced by it did not provide genuine independent evidence for Copernicanism (and added little over the following hundred years or so, as far as the technically proficient were concerned, that was more convincing than what was to be shown/seen in Kepler's tables). It provided powerful propaganda for the emerging traditions.[2] That evidence *could perhaps* have been rendered compatible with the Ptolemaic system, but the heart had gone out of that system, and the effort required to square Ptolemaic astronomy with the new evidence often made it seem easier just to go over to Copernicanism. Opposition to Copernicanism did not simply collapse, but persisted, and only gradually faded away over the 150 years after the death of Galileo in 1642.

In *The Copernican Revolution* Kuhn has thus tried to demonstrate three major 'incongruities' in the case of Copernicus:

- that such narrowly focused mathematical preoccupations could and did have such far-reaching scientific and even sociocultural consequences (143);
- that the Copernican revolution as we know it is hardly to be found in Copernicus' own writings (155); and
- that Copernicus' work could not have triumphed over Ptolemy's on the grounds that it was more accurate, more consistent, and less unnecessarily complex since it was itself inaccurate, overcomplicated and just as inconsistent (171).

The book accessibly provides Kuhn's first in-depth account of how the image of science by which most extant history, philosophy and popularization of science are possessed is badly off, in so far as these three incongruities (and more besides) are not reckoned with. The quantum revolution book, written nearly a generation later, and after the fame of *Structure*, but forbiddingly technical and austere, seems on casual inspection to be very different. Is it?

The Quantum case

More revisionist history. This time, it is meant to revalue a prevailing judgement on one individual, in this case, Max Planck. According to, for example, Isaac Asimov's popular history of physics,

> Planck's quantum theory, announced in 1900, proved to be a watershed in the history of physics. All physical theory that did not take

quanta into account, but assumed energy to be continuous, is sometimes lumped together as classical physics, whereas physical theory
that does take quanta into effect is modern physics, with 1900 the
convenient dividing point.[3]

This states just the position Kuhn is out to refute: 1900 *was not*
the dividing point between 'classical' and 'quantum' physics, and
Planck's theory did not introduce the idea that energy is discontinuous. The story is, in many ways, the same as that of Copernicus,
of a scientist who sought to solve a specific problem *and failed to do
so*, but who, in the attempt to find a solution, was driven by technical necessities to resort to a particular step – the introduction of
an idea, 'the quantum' – which he did not himself take fully seriously, regarding it as a temporary solution to a still unresolved difficulty. It fell to others subsequently to take the steps that would
introduce the *modern* idea of the quantum into physics, an idea
which Planck himself found it hard to accept. In this case, though,
and probably largely because of the much greater degree of
specialization, mathematicization, and use of technology in high
science 300 years and more after Copernicus, any emphasis on the
wider cultural environment of Planck's thought is diminished, and
attention is entirely focused on the course of Planck's own strug-
gles with his problem, and of the scientific materials that he drew
on in grappling with it, as well as relevant developments in the
wider science. This, at least, is the matter of the first main part of
Kuhn's study, the second part being concerned with the aftermath
of Planck's 'black body' work so as to show how, in physics, the
quantum idea actually emerged and took hold, and how Planck's
later work demonstrates his difficulties with the idea of the
quantum. The credit for its introduction, if one has to choose one
person to give the laurel to, goes to Albert Einstein.

Kuhn offers this dilemma for the prevailing view of Planck's
achievement: *either* one can hold that the idea of the quantum *was*
the one present in Planck's work of 1900, *or* one can accept that it
was not present and that this idea was later introduced by Einstein.
In either case, Planck must lose credit for its introduction, for if it *was*
present in that work, then this would be an idea that Planck did not
originate, but which he picked up through the work of a distinguished predecessor, Boltzmann.

The black-body researches were the means Planck used to
come to grips with what he regarded as the real problem, that of
Maxwell's demon (shortly to be explained). In accord with Kuhn's

general thinking, Planck was not seeking fundamental novelty and, like Copernicus, was setting out to take measures which would sustain, not challenge, classical physics. Planck expected that Maxwell's demon could be reconciled with classical physics, that the 'threat' it posed was only apparent, a purely theoretical one, and that it would therefore eventually, if not easily, be shown to be compatible with the classical framework.

When Kuhn published, some thirteen years after *SSR*, his book on the topic, *Black-Body Theory and the Quantum Discontinuity, 1894–1912* (*BB*), many of those who had enthusiastically awaited it were disappointed. The book's contents certainly looked as dry-as-dust as its title, and the book is relentlessly involved with pretty technical physics in a fashion very different even from *CR*. Anyone looking for lively and accessible polemics to follow up *SSR* would not find them, nor would they even find the *jargon* of *SSR*. The key words are absent. However, *all the ideas* from *SSR* are present (even if Kuhn had to write a paper just five years after the book to explain how this was so). We mention this now because it should be clear, given our account above of *SSR*, that Maxwell's demon was or at least can now be sensibly regarded as *an anomaly*, as an occasion for research, and just possibly for retooling. Here is a respectable piece of scientific work which does not work out the way that, in the framework of classical physics, *it should*. It provides Planck with a challenging opportunity for *puzzle-solving*. Though there must be something wrong with Maxwell's arguments, Planck cannot see what this is. The problem does not require the rethinking of classical physics, for the difficulty it represents seems very inconsequential to the physics, but Planck is not content to set the worry aside: he wants to dispose of it in a *principled* way. And this is what he tries *but fails* to do. Quantum physics originates in rejection of the idea that a principled reconciliation of classical physics and Maxwell's demon is possible. Thus, Maxwell's demon poses a highly technical puzzle that deeply troubles someone who is unhappy with the lack of a rigorous solution to a principled problem.

'Maxwell's demon' is a paradigm in the exemplar sense, a 'solved problem' . . . but one which offers a solution that is at odds with the prevailing paradigms, particularly the laws of entropy. The new (quantum) paradigm develops the idea that matter be considered discontinuous, conceived of in terms of discrete particles. It thus conflicts with the prevailing idea that matter is continuous. The idea that matter was made up of particles was itself one that developed within classical (that is, pre-quantum) physics, but in a way which kept it consistent with the idea that matter is also continuous, for while matter may be made up of particles, those particles themselves are embedded in the aether that fills the whole of space. Working out what the assumption of particulate matter might mean, Maxwell produces the kind of solution that is not acceptable in the contemporary physics for it is *statistical* in nature. It can, for Planck, be only a halfway house on the road to a proper explanation that will be given through 'mechanics', through the examination of the way constituents of matter interact with one another in a causal, strictly determined, non-statistical way (the assumption that this *kind* of explanation is the only proper, acceptable kind, being exemplary of the idea of paradigm, roughly in its disciplinary matrix aspect, in the deep or 'meta' commitments of a given discipline, or thought-collective). In the ideal case, knowing the initial positions of all the atoms composing some body, gas or liquid, assuming that body, gas or liquid to be isolated from all external influences, and knowing the laws governing the behaviour of those atoms, one could (in principle, at least) determinately predict all its subsequent states.

Treating heat as a matter of atomistic interactions, of the behaviour of molecules, Maxwell created a thought experiment: imagine a demon who regulated the exchange of molecules between two bodies, as though controlling a pipeline between them through which the molecules must pass from one body to the other. The internal composition of each body is to be understood as being composed of a distribution of differentially energetic molecules, some of which are 'hotter', that is, more energetic and faster-moving, than others – meaning that a liquid is not at uniform temperature throughout. The demon could, therefore, open and close the pipeline selectively so as to let through only the 'hotter' molecules from the colder body, while, reciprocally, only the 'coldest' molecules would pass from the hot body, meaning that the temperature of the hot body would rise, and that of the cold one fall, in contravention of the second law of thermodynamics, which insists that such a transfer of heat can *never* happen. Maxwell's thought experiment entails that because the number of 'hot' molecules in the cold body and of 'cold' ones in the hot body must be far fewer than the average in the respective cases, then the odds against this contravention occurring by natural processes and without the intervention of a demon must be very, very long indeed. However, extraordinarily long odds against is not the same as being ruled out as impossible by a law.

That the possibility the thought experiment allows is very remote meant that Planck regarded it as something that would never happen, and as resulting only from idealizations making up a limiting case. Still, the logic of Maxwell's argument from the idea of particulate matter looked sound, and so Planck did not just set the problem aside: there must be a non-statistical proof that it is wrong.

Planck set himself conditions under which the proof was to be achieved. First, the proof was to be given in terms of 'conservation laws' that then governed energy transfer and that are 'reversible' in the sense that it makes no difference to the laws in which direction a process is moving: the equations governing interactions 'look the same' whether two molecules are moving towards a collision with each other, or bouncing away from each other after colliding. Planck's ambition was to derive from these 'reversible' laws a proof that the process of heat transfer is 'irreversibly' from the hotter to the cooler body. Planck set himself the task of proving the irreversibility of entropic processes and the path he took became that of showing why the time reversible equations did not, in actual physical cases, apply in that way. The laws themselves were reversible, and so there was nothing intrinsic to them directly to rebut Maxwell's argument. This was the first of two constraints. The second was that he would make no 'special assumptions' about the constituents of matter; that, in other words, his proof would in no way depend on assumptions about the nature of molecular elements, other than that they were particulate. Thus, again in accord with Kuhn's ideas about normal science etc., Planck is setting technical and high standards, posing himself a very challenging puzzle.

The connection between Maxwell's demon and the black-body problem, as mentioned above, is the second law of thermodynamics. That famous law relates to the absorption and emission of light, and arises in connection with consideration of the quantitative relations between the temperature of a body and the amount of energy (including light) radiated by the body as it is heated. The 'black body' itself is a thought-experimental tool, proposed as a polar condition for understanding the relationship. The 'black body' would be one that would absorb all the light at all frequencies that fell upon it and would reflect no light at any frequency. If the body is rendered incandescent, then at that point it would release all the light it had absorbed in all of its frequencies. The problem was to work out a quantitative expression of the relationship between the distribution of light among its various features, and the temperature of the radiating body. It was in this connection that the black-body research became of interest to Planck, for if a black-body cavity (a closed body) containing energy waves is isolated from its environment, then a stationary state must set in, one which cannot be further changed without effects from the environment (and which cannot, therefore, be reversed, for that would require a change to start the reversal off). However, energy waves governed by Maxwell's equations would not of themselves reach an equilibrium, and it is only through the introduction of a thoeretical entity called a 'resonator' – since its role was to create resonations that could both absorb energy from the waves around it, and emit energy (in a transformed state) back into its environment – that the possibility of the equilibrium could be theoretically demonstrated. Planck needed the idea of the resonator to enable the adaptation of equations from mechanics to the continuous medium of air.

Incidentally, but decisively, it fell to Planck to prepare a deceased colleague's work on 'gas theory' for publication. The fact that gases were conceived as having an atomic (molecular) structure and were understood in terms of (determinist) mechanics offered resources for Planck's thinking about the black-body problem, drawing the statistical work of Ludwig Boltzmann to his attention.

Kuhn argues that it was Boltzmann's statistical work that led Planck to his quantum idea, but also misled him as to what he had actually done. Though Boltzmann's statistical ideas (bringing out the essentially probabilistic nature of the phenomenon) enabled Planck to persist in his (illusory) conviction that a reconciliation within the terms of classical physics could be found, Boltzmann himself expressed his statistical ideas in a classical and determinist way, thus obscuring from Planck the extent to which, using those statistical methods, he was operating inconsistently with his own other assumptions and (later) results.

Boltzmann (working out a formula for calculating the movement of molecules in a gas) computed in a way incongruous with his actual problem, calculating as if working from a non-statistical mechanics, in which the position of each molecule was initially known, and where the subsequent distribution could then be uniquely calculated. However, the initial distribution of molecules was not a real classical initial distribution, but a 'coarse grained one' (*BB*, 45). There was an infinite number of different ways in which the initial distribution of the molecules in a gas could be specified, and each of these would lead to a different outcome in calculating any later stage. Boltzmann's equations do not give actual values for particular molecules, but only average, or most probable, values. Boltzmann came to realize this, at least to some extent, but did not recast his way of stating things as if they gave fully classical, rather than purely probabilistic results. Boltzmann had conflated two levels of operation, and so simplified calculation, calculating the positions of molecules by treating them as being located in 'cells' (that is, volumes which can contain a large number of molecules), thus evading the need to specify the position of each individual molecule: 'only by moving subconsciously between these two ultimately independent conditions was Boltzmann able preserve for so long his predominantly deterministic way of discussing [his theorem]' (57).

He misled Planck, too, and it was not (Kuhn insists) until after 1906 that Planck realized that under quite usual physical conditions the distribution of molecules within cells affected calculations – for in Boltzmann's method, the need to make calculations about the distribution within cells was bypassed. Because they conflated average distributions with actual ones, both Boltzmann and Planck misinterpreted Boltzmann's methods as specifying actual, rather than most probable, distributions for molecules and therefore as tracing actual, rather than possible interactions between them (thus not seeing that the calculations were thoroughly statistical, not determinist at all).

By 1897, Planck was coming to realize that his electro-magnetic field developed on the basis of Boltzmann's example would not yield the non-statistical proof of irreversibility without special initial conditions (since that would be going back on the requirements initially set for a solution). In attempting to work out equations for energy waves in a black-body cavity, Planck found that his mathematics would not allow him to work out the equilibrium (that could only be disturbed by an external influence) for the black body considered as an empty volume. Planck introduced the 'resonators', imaginary devices which would receive and emit energy, to allow himself to work out the solution he needed. However, in 1898 he found himself again forced to go back on his initial requirements, now introducing a notion of 'natural radiation' to rule out possible configurations of microstates (for resonators in a black-body cavity) that would otherwise, like Maxwell's equations, allow for the reversibility of the tendency towards increasing entropy.

Having adopted a formulation – an entropy function – that barred some *statistically possible* micro-distributions as *physically impossible*, Planck thought he faced one outstanding obstacle to a full solution. He needed to prove that his solution was unique, that there was not a multitude, even an infinity, of other valid computations that would yield different results. Until 1899 he was convinced that the function was unique, but without strong proof. Planck's problem changed, under pressure of increasingly refined experimental results, from that of proving the irreversibility (as described above) towards that of articulating the radiation law that he had theoretically derived with results of actual experimentation, not just of thought-experiments. He could comparatively easily adjust his formula, but it would be ad hoc, leaving himself bereft of any strong derivation of the formula's results from theory (when the whole point was to provide a principled, not an ad hoc, derivation).

Planck's introduction of resonators into the black-body space had been criticized, and Planck came to see that it did present a problem. His procedure would only work for cases of more than one resonator. If the requirement that the atomic constituents were independent of each other was to be fully satisfied, then his proof should work for the single case (that is, the fact that there were or were not other atomic elements would not affect the behaviour of an isolated one). That his proof did not work for the single case meant the resonators in the multiple cases could not be considered independent, but this was inconsistent with his second constraint, that he would make no 'special assumptions' about the elements involved.

Rethinking his position, Planck needed to consider how the total energy available to the resonators in the black body was distributed among them. We now come to the point where, if anywhere, Planck introduced the idea of the quantum. His way of working out how the total energy (E) was to be distributed among the resonators drew on Boltzmann's statistical methods. To apply this method (whose details need not concern us) the energy continuum is subdivided into units of finite size, but the size of the energy elements into which the continuum must be subdivided cannot be arbitrarily set as though they were picking out just any level on a continuum of possible frequencies. These needed to be fixed proportionately to the frequency of the energy waves. Through the identification of 'h', a constant, later known as Planck's constant, and frequency (v), Planck contrived the energy element hv, which element could be divided into total energy, e, to give the energy elements to be distributed over the resonators.

It is here, if anywhere, that Planck came closest to breaking with classical physics. He found his calculations of energy levels would take only integral values. These could not be distributed to all and any points across a continuum, but could only be allocated to certain discrete levels. Kuhn maintains that Planck has been misunderstood as carrying out his work in the same way as H. A. Lorenz. Working rather later, Lorenz did indeed use an unequivocally quantized conception – that is, one which says that the energy levels themselves are discontinuous – but the whole point for Kuhn is that this conception is not present in Planck's work and has been read back into it. The fact that Planck's underlying concept was not the same as Lorenz's was obscured by Planck's adoption of a shortcut to avoid having to work out the distribution over resonators for all frequency levels, not just for one, this being a very cumbersome exercise. Thus Planck, as well as his readers, remained unaware that the actual problem he had solved was not the same as the one he was trying to solve. Instead of distributing the energy over a totality of resonators vibrating at different frequencies, Planck had distributed total energy over a number of resonators at a single frequency. Neither Planck's contemporaries, nor his subsequent readers had, therefore, previously been able properly to clarify Planck's conceptual reasoning at this juncture.

That the energy levels are discontinuous seems to be the quantum idea itself: why, then, does Kuhn doubt it?

The core of Kuhn's case is that the claim that the results of e = hv were restricted to integrals did not represent the post-classical quantum. Kuhn admits there are passages where it seems that Planck does indeed literally use the modern quantum idea. However, Kuhn tells us, in a most unusual manoeuvre, 'fortunately for the consistency of Planck's thought these passages need not be read literally' (128). In view of Planck's general and persistently classical-in-nature way of thought, and in light of Planck's lectures of 1906, there are grounds for insisting 'that they should not be' (128). Kuhn's way of understanding Planck is the first to render his (Planck's) own intellectual development . . . continuous, consistent.

Kuhn holds that Planck was not really giving up on the conception of the energy spectrum as a continuum in favour of it as a series of discrete states, but was only attempting to simplify his statistical computations by describing that continuum in discrete terms (130). For Planck the relationship e = hv might entail the sub-division of energy levels in integral terms, but it remained a means of subdividing a continuum. At the same time, however, it also manifested a truth about the world. The facts of nature allowed only these energy levels, that was sure enough; but this could not, for Planck, be the final position. There remained a need to understand how continuous nature could yield discontinuous results, and Planck hoped that this would be resolved by better understanding of microscopic energy emission processes, and thus, through the newly developing electron theory (which would give physical reasons why the values were discrete even though the phenomena were not).

Planck's hopes could not have been fulfilled by electron theory, Kuhn claims, but Planck could maintain the illusion that there could be a classical solution to the radiation problem, the hope being sustained by errors in his own reasoning about the distribution of the resonators across the cell divisions in the volume of the black-body cavity that allowed for only some of the locations within the boundaries of a cell to be available to his resonators.

There is much more in *BB* than this re-examination of the logic of Planck's own work, as Kuhn seeks to show both how the modern idea of the quantum originated *after 1901* (when Planck had (temporarily) abandoned work on black-body radiation) and to show that Planck had difficulty in ingesting that idea.

It was two physicists, Albert Einstein and Paul Ehrenfest, working independently of each other, the latter critically reworking Planck's black-body material, who brought in the idea of the quantum as we know it. Einstein provided a physical basis for Planck's results: arriving at the requirement through his own independent line of inquiry, initiated and pursued without reference to Planck's black-body work, Einstein saw that Planck's theory, 'properly understood', required Einstein's own light-particle hypothesis to explain radiation in the higher frequencies. Einstein argued that high frequency radiations behave like a collection of particles, and his formula for the energy of these particles turned out to be the same as Planck's energy elements. This was the basis for Einstein's reinterpretation of Planck's theory, which, he concluded, had not been compatible with classical theories, and required a break with them, though 'Einstein immediately insists that Planck himself had not noted the need for such a break' (*BB*, 185). Einstein famously remarked that the fact that beer is sold in litre bottles does not mean that beer exists only in litre quantities:

Planck's equation allows only discrete quantities of energy, but this does not imply that energy exists in discrete quantities.

Einstein and Ehrenfest had devised the quantum concept, but it was not so obvious to other physicists that they *instantly* took it up. Though Einstein was rapidly becoming influential, his prestige was not then such as to convince many of the merits of his light-particle hypothesis, and the association with this of the quantum idea actually discouraged acceptance of the latter too. 'If the physics profession was to recognize the challenge of Planck's law, better established figures would need to be persuaded that it demanded a break with classical physics' (189). One such was H. A. Lorenz, who came to see that the idea of an elementary quantum of energy might serve to make sense of Planck's formula, but Planck's theory, as it stood, seemed to require that his resonators absorb and emit energy in an entirely continuous way. Questioning Planck on this point, he drew from the latter his 'first known concession of the need to restrict resonator energy' (194). Lorenz was already a considerable figure, with acknowledged expertise in the relevant field, but his version of the arguments also avoided certain difficulties which afflicted those of Ehrenfest and Einstein. By conclusively eliminating the previously proposed mechanisms for redistributing frequencies – namely, moving resonators and colliding molecules – he made electron theory seem to be the only basis for deriving Planck's law.

Having accepted the need to restrict resonator energy levels, Planck remained reluctant to admit the necessity for more than a minimal break with classical physics, though by 1910 he had *'at last'* become 'firmly and publicly committed to the entry of discontinuity and the abandonment of some part of classical theory' (200).

From 1907 and increasingly after 1910, the quantum idea took off, and both the number of physicists working on it and the number of areas they were working in expanded, while interest in blackbody researches themselves declined. Kuhn avers: 'During the years 1906–1910, back-body theory was rapidly taking the form Planck was once thought to have given it in 1900' (205). Planck reluctantly came to terms with post-classical ideas *to some extent*, attempting to incorporate the new developments with respect to the quantum into his own thought, producing a second theory of radiation. Kuhn's account is again revisionist. Planck's contemporaries and historians have seen in this later work a retrograde conservatism by comparison with his earlier, turn of the century, revolutionary radicalism, where Kuhn sees the personal radicalism of a conservatively inclined individual. Planck is only (after 1910) attempting to take the steps he had avoided but that had now been taken by others and imposed themselves upon him, giving him 'a new perception of his achievement in the area he had made his own, the electromagnetic theory of black-body radiation' (311).

Forced to admit discontinuity, Planck still did not easily accept the idea that energy did not comprise a continuum. First, he sought to incorporate the discontinuity by proposing a threshold level for absorption of radiation, then revised that into the view that the discontinuities were in emission rather than absorption. Scientists seeking to understand the nature of the quantum were often influenced by Planck's theory, at least until Nils Bohr's work on spectra showed that 'an identical process was at work in absorption and in emission' (252), and thus precipitated the second theory's gradual disappearance.

Though taking away Planck's status as the founder of quantum physics, Kuhn nonetheless deems Planck's two ultimately unsuccessful theories of radiation to have been consequential, playing an important role in its creation, giving others the confidence to proceed in the new direction going beyond Planck's own ideas. It was, therefore, a collaborative effort, an *extended, not an instant* revolution (as usual). Planck's theories had not 'called for the existence of a discrete energy spectrum, the characteristic that in retrospect has seemed the essential characteristic of quantum theory', but Planck had still been a primary contributor to the development of a theory that (ironically) 'he never came quite to believe' (254).

Though the terminology of *SSR* is absent from *BB*, it should be clear that all the ideas are present, and that this work, together with *CR*, provides clear substantiation of the arguments of *SSR*. If what we say is the case, why did Kuhn *avoid* his own vocabulary? For two main reasons, we think: to avoid the kinds of misunderstandings that he had so often encountered of words such as 'paradigm', and, more generally, to keep himself out of a straitjacket. But Kuhn *wasn't* happy with the general presumption that *BB* had nothing to do with or even contradicted *SSR*. Within five years of publishing the book, Kuhn had to write a paper, republished in the second edition of the book, to point out the connections with *SSR* and to explain himself. He does not want to jettison the ideas of *SSR*, though neither does he want them to become a set of rigid preconceptions, a theory to be imposed on the historical data:

> I do my best, for urgent reasons, not to think in these terms when I do history, and I avoid the corresponding vocabulary when present-

ing my results. It is too easy to constrain historical evidence within a predetermined mold. . . . Often I do not know for some time after my historical work is completed the respects in which it does and does not fit *Structure*. (363)

Planck's 'conservatism' was not some reactionary posture, but, as Kuhn's whole book seeks to show us, a deeply reasoned and entirely reasonable response to an intriguing, and at the time purely technical, challenge, one that, so far as he could see, had to be dealt with by being disposed of rather than accepted on its own (apparent) terms. Maxwell's demon and, subsequently, the experimental results on cavity radiation, with the variation in the release of high frequency light, provided classic and, to Planck, nagging puzzles.

In Kuhn's account his motivation is very plainly: puzzle-solving. Planck was trying to get a satisfactory solution to unexpected anomalies, within the restrictions of his requirement to avoid special hypotheses, and of forming the solution out of 'conservation' laws: showing that essentially reversible equations could nonetheless be used to yield a proof of irreversibility; attempting to adapt techniques developed in the theory of gases to the case of thermodynamics; working on the problem theoretically, but responding to experimental evidence, and to the objections of critics; trying determinedly, and over several years, and by different means to reach this solution. Planck was a normal scientist – a beautifully ambiguous sentence, true in both its meanings, after reading Kuhn. That is, Planck was a normal scientist, one like any other, just a brighter one who could be said to have helped the community 'sleepwalk' into a revolution; but he was attempting to practise normal science too, assuming that the problem of Maxwell's demon must have a solution within the prevailing paradigms. Kuhn thus shows that the actual concrete work of normal science is not something starkly contrastive with revolutionary science, but something that can merge into it. Any 'critics' of Kuhn's supposed 'dichotomy' between normal and extraordinary science should read his corpus better before taking themselves to have refuted Kuhn.

So, like Copernicus, Planck was judged by Kuhn to have been to a greater extent than previously realized a failure – he had not solved his actual problem, had not succeeded in proving the irreversibility of the law of entropy in terms of conservation laws, and had, indeed, only been satisfied with his achievements on the basis of a misconception about what they were. Again like Copernicus, the revolution Planck has/had been credited with, and one that had been named on the basis of the term that he had introduced – the quantum revolution – was largely not his achievement at all, but that of those who came after him. The use of the same term, 'quantum', had fatefully misled both contemporary scientists and subsequent historians into supposing that it signified the same idea.

Bohr's atom

Also worthy of mention as a kind of 'supplement' to the black-body book is the account, written by Kuhn with John Heilbron in 1969,[4] of the development between 1911 and 1916 of Nils Bohr's conception of atomic structure. Like the black-body study this is a relentlessly technical reconstruction of the way in which Bohr eventually worked out his conception. Like the black-body study, it means to be revisionist, though rather more mildly so, principally drawing attention to a neglected connection between Bohr's work and that of another physicist, J. W. Nicholson. However, its broader purpose is to fill out the story of how Bohr developed his ideas, and some of this is speculative, reconstructing what Bohr must have thought and reasoned at times when there is no record of what he was doing. The study is worth mentioning because this is a period of crisis in physics, with the quantum *revolution* taking place. Bohr differs from Copernicus and Planck in that unusually he *is* looking to make fundamental change, convinced by the state of things in physics that the laws of mechanics thought capable of governing the behaviour of atoms would not be able to do so. They would break down in this area, and a quantum view would be needed. It is precisely in revolutionary periods that scientists can meaningfully aspire to fundamental novelty, and may be 'casting about' for radical solutions.

.However, *even so*, Heilbron and Kuhn insist that the answer to why Bohr's work followed the precise course it did is not to be found in his 'general conviction of the need for quantum theory which Bohr drew from his thesis research, but rather in certain *specific* problems with which he busied himself until the end of his year in England. They helped direct his reading and uniquely prepared him to recognize the special potential of the nuclear atom.'[5] Thus, the study is of the determination of the technical problems in physics that Bohr grappled with, and the exigencies that drew his attention to them or contributed to his thinking about them – such exigencies as his inability to get on with particular scientists, his choice of laboratory to work in, and the fact that he came across physics papers that made an important difference to him. In other words, the story is mainly about how Bohr's inclination towards quantum theory was given *technical bite* – though it did not, in his case as in others, mean that his achievement was a solution to the problems that he had initially been trying to solve, though it was to issue in revolutionary results: again, detailed puzzle-solving. The case exhibits the difference – in the details, crucially – between Planck's basically classical conception of the quantum and Bohr's fully quantized one:

> As Bohr develops the limiting case, it is apparent that classical and quantum computations coincide only in their results; *the models and the mechanisms of radiation remain distinct*. Planck's oscillator, in contrast, could with relative ease be viewed as itself behaving classically in the low frequency limit.[6]

Conclusion: implications

Kuhn's historical studies reiterate central theses from *SSR*. First, that the comparative relations between two rival paradigms, *at the time they are in vigorous competition*, may be, for those involved, anything but clear-cut. Setting aside presuppositions about what has happened since these two great revolutions, one can come to see that the now-outmoded doctrines were much less *obviously* mistaken than they may now be taken to be, and that the line between adherents of the old and the new paradigm is not so easy to draw as, in retrospect, it seems: *both* Copernicus and Planck have been misunderstood as revolutionary heroes when their motivation was to shore up, not cut down, the paradigm in place. Both studies support the idea that the search for fundamental novelty is *not* what is always, or even usually, involved in scientific innovation.

Further, notice that Kuhn claims that the victory of Copernicanism was no knock-out result but more a victory through attrition, and also suggests that the Ptolemaic tradition could have been shored up against the Copernican challenge, but that eventually there was no will to do this: it had been worn down rather than *decisively refuted*.

One sometimes hears grumbles that Kuhn didn't really free himself of the hero-worship of Whig history (the worship of past heroes, and of ourselves as the *telos* of those heroes), because he spent his time talking about scientific revolutions, not giving us accounts, as many in Science Studies now do, of the nitty-gritty of normal science. But these grumbles are misplaced and misleading. Both *CR* and *BB* may be studies of scientific revolutions, but they are, in fact, mostly studies of *normal science*. There will be more discussion about normal science below, but these two studies should militate against the idea that 'revolutionary science' is something that is done with a different *motivation* than 'normal science', for with both Copernicus and Planck their 'revolutionary' contributions resulted from what were unquestionably exercises in puzzle-solving, of the kind, one might even add, that surely only appeal to someone with a 'geekish' nature. It should explode, too, any idea that normal science puzzles are intrinsically trite, or that there are easily discernible relations between the apparent and actual significance of any given problem.[7]

Part II

Critical Issues

We divide the critical issues arising from Kuhn's work under three broad headings. First, those associated with the concepts of normal and revolutionary science. These questions emerge naturally from our discussions in part I but deserve further detailed consideration here. We believe that the concept of 'normal science' is easily misunderstood and that properly understood it is what makes Kuhn's contribution to *philosophy of* science revolutionary.

For Kuhn virtually all science is normal science, and normal science is a paradigm of rationality. On our account, it is 'normal science' not 'revolutionary science' that 'wears the trousers', and if it is thought to be the other way around, this distorts the understanding both of what Kuhn has to say and of the natural science he is talking about. It also gives a wrong impression of what Kuhn might mean for other intellectual domains than the natural sciences.

Kuhn is contrasted with Popper in the general direction of his thought, and we argue that the substantive differences between them need not be quite so divisive as they may seem to be. We then go on to discuss how Kuhn's philosophy of natural science is and is not to be taken as a philosophy of 'social science'.

In the second section of part II we try to deal with what is, for many, the sticking point with Kuhn, his idea of incommensurability. Though the idea is clearly essential to Kuhn's thought, it seems to many to be a fatal flaw. It seems to lead directly to undesirable consequences:

that nature has no role in determining the nature of scientific
    thought;
therefore to relativist or idealist implications; and
thereby to Kuhn's entrapment in a performative contradiction,
    that is, telling us we cannot understand previous science by
    showing us the right way to understand previous science.

The section on incommensurability will focus on the idea of translation because this is how Kuhn and his most distinguished critics
have cast the problem. Can there be translation between different
scientific paradigmatic theories? Does Kuhn say not?

Here, as on other occasions before, it is wise to pause for a
moment and ask, what exactly is the question? Does Kuhn say that
'there can be no translation?' or does he say 'there can be no point-
for-point, word-for-word translation?' or what? The difference
between the two questions might not be instantly apparent, but it
is very important in understanding what Kuhn is getting at. He
does deny that there can be word-for-word, point-for-point translation – but that is merely reiterating his point about discontinuity
between successive paradigms.

Does incommensurability mean that people separated by history
or culture live in different realities – different 'worlds' as they are
usually called? In other words, is Kuhn, whatever he might say, con-
victed of Relativism, thinking that there is no external standard
against which different views can be compared, that one way of
looking at things is as good as any other?

3
Kuhn and the Methodologists of Science

PHILOSOPHY OF NATURAL SCIENCE

We return to the topic of normal science, and to the disputes between Kuhn and two of the other epochal figures in philosophy of science in the 1960s: Karl Popper and Paul Feyerabend. The bone of contention was the normative implication of Kuhn's idea of normal science. Both Popper and Feyerabend saw that implication as being *conservative*. Thus, there are two questions addressed in this chapter: Does Kuhn's philosophy of science lack a normative aspect? And if it possesses a normative aspect, is this conservative in nature? Answering these two questions will hinge on a further question: does either Popper or Feyerabend fully grasp the nature of Kuhn's concept of normal science and its subversive consequences for their own philosophies of science?

Philosophy of science commonly conceives of itself as a normative venture prescribing how scientists should proceed. Philosophy of science often attempted to identify a scientific method, but it was just this idea that the philosophers of science involved here were out to reject. Karl Popper had become, by the 1950s, the dominant figure in philosophy of science, and had gathered a school around him. Popper denied that the identification of a general scientific method was of any interest to philosophy of science: *how* people came up with their ideas has nothing to do with whether they are any good or not. Feyerabend and Kuhn attempted, in their different ways, to instigate a revolution against the Popperians, though both agreed that the philosophy of science was not about the identification of any scientific method. These two had very similar views on this: the scientific method was a historical fiction; no such thing was to be identified in the activities of actual scientists throughout history. What then might be the normative implications of philosophy of science?

Let us say that a key to understanding their disagreement is that both Popper and Feyerabend wanted to give philosophy of science a role which was *extrinsic* to science itself. Both were driven by anti-authoritarian motives. Popper was particularly incensed by the threat that totalitarian politics represented to freedom generally, and to the growth of science in particular. Thus, Popper insisted – in terms of a liberal politics – on the urgency of maintaining a free society, one with institutions that allowed the possibility of criticism, for it was that that he identified as the very essence of science. Reflecting generational differences, Feyerabend was less troubled by the menace of totalitarian regimes than by the post-atomic arrogance of scientific institutions. Rather than needing defence against authoritarians, the sciences themselves were becoming a serious authoritarian threat. Thus, the need was to encourage scientists to be less authoritarian in their science (and therefore outside it as well). Feyerabend struck an anarchist pose. Both understood Kuhn to present science as (rightly) authoritarian. It is not surprising, then, that both accused Kuhn of conservatism.

Popper's central, founding problem was the 'demarcation problem', the need to distinguish metaphysics from science (where there was a danger – as with Marxists and Freudians – of the two becoming confused). The difference, tied to the essentially critical nature of science, was that scientific hypotheses are constructed so that they allow the possibility of refutation. Metaphysics does not, in the way science does, make predictions that are sufficiently determinate to compare with the facts, and for discrepancies between the two to be identified. Popper's recommendation is that people should move out of irresolvable metaphysical disputes and into the resolvable problems of science by casting their thought in a form allowing the possibility of refutation. Popper wanted to change the whole philosophy of science around. It had been preoccupied with the question as to how scientific ideas can be confirmed. But, says Popper, they *cannot*. The only *conclusive* judgements which can be made on hypotheses are negative ones, that they have been disproved. No matter how much evidence is adduced in support of a genuine scientific hypothesis, this can never *conclusively* confirm the truth of the hypothesis.

A view of the nature of science is proposed: it is an essentially critical enterprise, exclusively devoted to making efforts to disprove theories. That scientists should put forward theories that have the potential for refutation is not perhaps something that can count as a *recommendation*, for, on Popper's terms, it is the fact that they are doing this that makes what they do scientific (not metaphysical). Feyerabend, rather, thinks science is not so much critical as authoritarian. It is increasingly *dogmatic*, both within itself and progressively throughout the society. It will brook no alternative to its point of view, and regularly exceeds its competence in asserting itself against practices (such as alternative medicines, astrology and the like) that it deems inferior. He sees this view reflected in Kuhn, but feels that the authoritarianism is *approved of* by Kuhn.

Popper rejects the classical idea of scientific method, but is, of course, a methodologist prepared to prescribe. Feyerabend is different: he sees a need to reduce authoritarianism in science itself. A central element in that authoritarianism is the insistence on a single, unified frame of thought that recognizes no valid alternatives (this sounds very much like a paradigm, does it not?). For Feyerabend, this is just a prejudice, and one that needs overcoming. Scientists should not assent in one dogmatic scheme, but should resist this hegemony by creating alternative scientific schemes, realizing plurality within any scientific field.

Normal science

It may seem odd that Kuhn's confrontations with Popper and Feyerabend centred on normal science. One might have expected that the concept of scientific revolution would be the more provocative one. Those who find in normal science a holdover of the old paradigm in the philosophy of science, a less exciting aspect of Kuhn's thought, may be mistaken.

'Normal science': on the surface an unexciting, even dreary, term. Why does Thomas Kuhn, seemingly the philosopher of scientific revolution, dwell on this boring-sounding topic? Surely scientific revolutions are the 'meat' of Kuhn's account, and normal science merely their adjunct? How can Kuhn, this revolutionary in the philosophy of science, allow, apparently, that science may progress normally, cumulatively, outside of revolutionary interruptions? If one looks, for instance, at the document which above all records the clash between Kuhn and his critics, the volume *Criticism and the*

Growth of Knowledge,[1] which comprises the key papers given when Kuhn met face to face with Watkins, Toulmin, Popper, Lakatos, Feyerabend and others in July 1965, one finds a different picture, as is obvious in the titles of some of the papers given there: Karl Popper's 'Normal science and its dangers', and even more starkly, J. W. N. Watkins: 'Against "normal science"'. Furthermore, Feyerabend, often (and often rightly) pictured as an ally of Kuhn's, had similar – and equally *drastic* – complaints:

> I was quite unable to agree with the *theory of science* that [Kuhn] himself proposed; and I was even less prepared to accept the general ideology which I thought formed the background of his thinking. This ideology, so it seemed to me, could only give comfort to the most narrowminded and the most conceited kind of specialism. It would tend to inhibit the advancement of knowledge [because its effect] is to restrict criticism, to reduce the number of comprehensive theories to one, and to create a normal science that has this one theory as its paradigm.[2]

Popper and Feyerabend saw Kuhn's work as unwelcome because it carried dangerous (conservative) implications for the practice of science in the idea of normal science. Popper thought normal science meant that scientists could and should be cautious, while Feyerabend feared that Kuhn encouraged scientists to be obedient to one paradigm, thus reinforcing science's authoritarianism. Here is Feyerabend being quite direct on the matter, in a letter to Kuhn which sounds unmistakable notes of warning, and of negativity:

> You thereby take your readers in . . . I do not object to your [having the] belief that once a paradigm has been found a scientist should not waste his time looking for alternatives but try working it out . . . What I do object to most emphatically is the way you present this belief of yours; you present it not as a *demand*, but as something that is an obvious consequence of historical facts.[3]

And here is a more indirect but nevertheless significantly anti-Kuhnian remark, also from the early 1960s:

> It is very important nowadays to defend . . . a normative interpreta-tion of scientific method . . . even if actual scientific practice should proceed along completely different lines. It is important because many contemporary philosophers of science seem to see their task in a very different light. For them actual scientific practice is the mate-

rial from which they start, and a methodology is considered reasonable only to the extent to which it mirrors such practice.[4]

For 'actual', one can (we think) read 'normal'.

However, the key question is not whether Kuhn offers this kind
of threat, but whether these critics have understood what normal
science is, and whether they have thought through the extent to
which *in practice* they must, willy-nilly, agree with Kuhn.

What has upset Paul Feyerabend so much, prompting such a
colourful attack on his 'ally', and what has incensed the orthodox
Popperians?[5] A partial answer is provided by Barry Barnes:

> It was [his] account of scientific revolutions, with its explicit
> Relativistic implications, which ensured that Kuhn's work became
> widely known. Normal science was not so arresting a phenomenon,
> and some commentators, noting Kuhn's references to its 'cumulative'
> character, even managed to interpret it as nothing more than 'ratio-
> nal enquiry' in the traditional sense. The consequent image of revo-
> lutions as impassable crevasses ripping across the path of rational
> scientific progress was vivid and exciting, and aroused great interest.
> This was, however, an image sustained entirely by an outmoded and
> untenable stereotype of the growth of knowledge; and when it is
> set aside the value of Kuhn's concept becomes more open to
> question, and several weaknesses become evident in the manner of
> its formulation.[6]

Why does normal science arouse the Popperians and
Feyerabend; and how it is that *'normal* science' can be reasonably
asserted by an influential social theorist like Barry Barnes to be
Kuhn's *fundamental* theoretical contribution?

What is normal science?

Normal science is, as we read Kuhn, *very nearly all* that scientists
actually do. It is a process of extending and filling out the realm of
the known; it does not look for fundamental novelties. None the
less it is far from the kind of routinized and intellectually empty
drudgery some commentators assume. The tasks of normal science
vary enormously, and even the most ordinary of them can be
immensely challenging.[7] And it is, normally, entirely conducted
within the ambit of a single paradigm.[8]

Kuhn writes of five foci for factual (experimental) scientific investigation. These are:

1   *The further investigation of 'that class of facts that the paradigm has shown to be particularly revealing of things'* (SSR, 24)   Tycho Brahe achieved fame as a great astronomer on the basis of his detailed and systematic observations of those features of the solar system and the stars suggested as crucial by Copernicus. Tycho made no great discovery or theoretical innovation, but 'the precision, reliability, and scope of the methods [he] developed for the redetermination of a previously known sort of fact' (*SSR*, 25) made his reputation. Kuhn thinks that a significant proportion of normal science consists of this kind of activity.

2   *The attempt to demonstrate the agreement of the paradigm with the world, at the few places where one might claim that they can be directly compared*   An example of an explicitly anti-Popperian idea (Popper would exclaim that scientists ought not to be trying to confirm a paradigm's effectiveness) is the creation of special telescopes designed to demonstrate the presence of the annual parallax of the stars predicted by Copernicus.

3   *The determination of the value of natural constants*   For instance, the working out of Avogadro's number, of the universal gravitational constant predicted by Newton, and of the speed of light.

4   *The determination of quantitative laws*   For instance, the determination of Boyle's Law, or of Joule's; and sometimes of their subsidiary versions in more specific or anomalous circumstances. Kuhn writes: 'Perhaps it is not apparent that a paradigm is prerequisite to the discovery of laws like these. We often hear that they are found by examining measurements undertaken for their own sake and without theoretical commitment. But history offers no support for so excessively Baconian a method. Boyle's experiments were not conceivable (and if conceived would have received a different interpretation or none at all) until air was recognized as an elastic fluid to which all the elaborate concepts of hydrostatics could be applied' (*SSR*, 28). And so on.

5   *'Alternative ways of applying [a] paradigm to new areas of interest'* (SSR, 29)   These are the only aspects of normal science – of the vast *vast* majority of science as it is actually, and not unwisely, practised, according to Kuhn – that *remotely* resem-

ble the 'paradigm' one may previously have had in mind of science as exploration, as discovery, as groundbreaking, etc.

Then there are theoretical problems of normal science that fall into nearly the same classes as the experimental and observational (*SSR*, 30). It's not surprising that Kuhn should suggest this kind of symmetry, because part of his ubiquitous point is that there tends throughout the sciences to be far more of an integration of the experimental and the theoretical than abstract theoreticist or empiricist models of science would have us believe. For Kuhn, the experimental can only be intelligibly planned or marshalled in the service of theories – and theories will not get off the ground without being instantiated in exemplars, in equipment, etc. You can of course, also and importantly, have thought-experiments – which Kuhn thinks never have a purely 'conceptual' import, but always involve 'learning about the world as well as about the concept' (*ET*, 258).

Kuhn sums up by saying:

> These . . . classes of problems – determination of significant fact, matching of facts with theory, and articulation of theory – exhaust, I think, the literature of normal science, both empirical and theoretical. They do not, of course, quite exhaust the entire literature of science. There are also extraordinary problems . . . [but these] emerge only on special occasions prepared by the advance of normal research. Inevitably, therefore, the overwhelming majority of the problems undertaken by even the very best scientists usually fall into one of the . . . categories outlined above. (*SSR*, 34)

Thus Kuhn thinks of normal science as puzzle-solving, rather than as like the creation or discovery of an entirely new puzzle. Almost all science, almost all the time, is not mediocre simply because it is like the ingenious figuring out of how things fit together in a vast and extremely complex puzzle. The places for almost all the pieces may have been prepared ahead of time, but this doesn't make the puzzling out of how and where they fit together any less difficult.

First of all, the issue is not whether this kind of characterization makes science interesting from the point of view of the philosopher, and whether philosophers (or indeed Kuhn) find this idea of normal science unappealing, unexciting and dull. What matters in Kuhn's account – and this is a vital 'sociological' point – is whether scientists find these kinds of things interesting enough to fill up their careers, even their entire waking lives? Kuhn's suggestion is that scientists *do* find these things interesting and satisfying, and give each other prestige and honorifics on their basis; it is not only mould-breaking paradigm shifts that achieve high scientific recognition. Further, the substantiation of Kuhn's claim would involve testing out his claim that there is little else to be found within the scientific literature except for these kinds of attainments. When one thinks of the relatively few examples that serve the historians and philosophers of science to illustrate their arguments, and then reflects on the endless rows of scientific periodicals filling the science sections and periodical stores of the libraries, one can see that a tiny proportion of the work done is memorable in the popular sense. The great mass of it will be forgotten, even in its own discipline, but that is (as Robert Merton tried to point out to ancestor-worshipping sociologists) one of the big differences between natural and social sciences: the natural sciences move on. However, the fact that scientific work is not publicly notable outside its (quite small) field,[9] and that it is subsequently abandoned or anonymously absorbed into new paradigms does not mean that it must be worthless – it is the essential stepping stone to further scientific work.

It is only very rarely, in revolutions, that the entire puzzle can be overturned and replaced by another. And of course, one wouldn't normally expect the pieces of the old puzzle to fit into the new one.

Unsupported by evidence?

Let us first address an important point which we have not, so far, explicitly and fully addressed, namely our insistence that Kuhn's objectives are overwhelmingly philosophical. It might seem – indeed it does seem to some 'sociologists' – that the implication of what Kuhn says, whether he acknowledges it or not, is that the problems of philosophy of science should be handed over to sociology (of science). (We put the quotation marks around the term 'sociologist', because we see problems with the demarcation of sociology (as purportedly an empirical pursuit) from what we might call a 'philosophical sociology' or, alternatively, 'a philosophical anthropology', more philosophical than empirical in nature. The difference is not an uncomplicated distinction between philosophy (done by people known by their professional title as 'philosophers'), on the one hand, and empirical science (done by people professionally identified as 'sociologists') on the other. There is a problem in sociology's identity, such that it can be argued (as Peter Winch did) that it announces itself as an empirical discipline but is overwhelmingly entangled with philosophical issues. For a simple example, note the way in which the Edinburgh school of the strong programme in the sociology of science – led by David Bloor and Barry Barnes – was programmatically insistent that philosophy of science must give way to sociology of science. But when requiring a theory of learning to advance their arguments, they did not turn to empirical studies of child rearing, but adopted Quine's unquestionably philosophical and highly controversial philosophical reconstruction of what concept acquisition in children must be.) Of course, there is in this a difference between the question of whether the implication of Kuhn's work is that the chalice should pass from philosophy to sociology, and the one of whether he intended any such implication? If the question under consideration is the latter, then the answer to that at least is that he unequivocally didn't! Indeed, he later came explicitly to disavow the whole idea.

That there is seemingly astonishingly little empirical (sociological) evidence or data offered anywhere in the corpus of Kuhn's work to support his picture of the nature of (normal) science of the sort that shows just what proportion of scientists are doing what sorts of things clearly indicates that Kuhn's interest was not in providing empirical generalities. There are of course the historical case studies, but one could still object that Kuhn has not given us explicit accounts of humdrum (non-famous) normal scientific endeavours. Kuhn clearly had more of a grasp than almost any other enquirers into the philosophy of scientific change (into the theory of scientific methodology) of the actualities of the pursuit of science; but it is surprising in an academic treatment of the matter that we have mostly simply to take this on trust from him. Moreover, it is yet more surprising that this fact has elicited so little comment down the years, given that Kuhn's influence on the social sciences has been so great. Why do sociologists not demand of Kuhn at least as high a standard as they would demand from any piece of work submitted to one of their own professional journals? (Part of the answer may lie in the disguisedly philosophical nature of 'social sciences' suggested above.)

Even Barnes, while noting in passing (without complaint or suggestion for follow-up) the complete lack of empirical support offered by Kuhn for his suggestive account of the nature of scientific training, doesn't bother to mention that Kuhn's entire account of the practice of normal science – his 'fundamental' contribution – is largely unevidenced.

Let us mention just a couple of examples (a few of the many possible), taken from Kuhn's discussions of normal science, of what we are talking about:

A part of normal (scientific) theoretical work, though only a small part, consists simply in the use of existing theory to predict information of intrinsic value. The manufacture of astronomical ephemerides, the computation of lens characteristics and the production of radio propagation curves are examples of problems of this sort. Scientists, however, generally regard them as hack work to be relegated to engineers or technicians (*SSR*, 30).

Is Kuhn in danger of falling victim to the 'great scientist syndrome' which as a matter of general policy he attempts to resist? In order to answer this question, it would be useful to be shown the evidence that scientists *do* generally regard these things as 'hack work'. Kuhn offers none, here or elsewhere.

Kuhn again:

> The scientific enterprise as a whole does from time to time prove useful, open up new territory, display order, and test long-accepted belief. Nevertheless, the individual engaged on a normal research problem is almost never doing any one of these things. Once engaged, his motivation is of a rather different sort. What then challenges him is the conviction that, if only he is skillful enough, he will succeed in solving a problem that no one before has solved or solved so well. Many of the greatest scientific minds have devoted all of their professional attention to demanding puzzles of this sort. (*SSR*, 38)

Here we have a truly central claim of Kuhn's. But where is the 'evidence' for it? Where is the sociology/anthropology?[10] Kuhn indirectly supports his claim a little elsewhere through historical remarks about the science of (for example) Roentgen; but precious little. And to offer no empirical support at all in the first instance for a vital contention such as that normally scientists are almost never displaying order or testing long-accepted beliefs – is this not a remarkable procedure?

Is it not especially remarkable, given that Kuhn early in his book claims explicitly (and in line with his own groundbreaking suggestions as to the true nature of scientific pedagogy) that he will introduce his readers to 'examples [paradigms?] of normal science or of paradigms in operation' prior to much later giving a 'more abstract discussion' of them?[11]

To make sure we are not being unfair to him: Kuhn at least makes a start at rectifying the omission, if such it is, in the second edition of *SSR* (176). We should not perhaps expect *too* much from Kuhn on this score. He is not a sociologist, and in the final analysis has little interest in the actual nature of professional (thought-)communities, etc. His interest, one might say, is rather to use reasonable sociological guesses suggestively against dubious philosophical claims and overgeneralizations.

In sum: there may be good reason to believe Kuhn's very general claim(s), but it is troubling to have to rely on his say-so. A good question, we think, to ask oneself when presented with a bold set of unevidenced abstractions such as those quoted above is always this: *What would it take to make belief in what Kuhn is saying here reasonable?* Supplementarily: What data would be helpful? What studies would it be nice to know that Kuhn conducted, or at least read, to support his claims (if they are meant to constitute more than provocations against the complete abstractions offered in the accounts of scientific theory verification and falsification he was rebelling against)? What would it take to 'confirm' Kuhn's philosophy of science here?

And so, now, do we take any of the above as reflecting negatively on Kuhn, or at least on the claim to find in Kuhnian 'normal science' the centrepiece of his revolution in the philosophy of science? As a matter of fact: not at all. Rather, we take it as a strong indication that one of our central claims is true: that Kuhn is a *philosopher* of science – that it is in his philosophical claims and provocations and suggestions that his real interest lies. For sure, his home discipline was history of science, and within it he made valuable contributions. In a sense, in their detail, they are of narrow interest (they will never be widely read). As we have seen, they have deep import in providing Kuhn with examples to exemplify and dramatize the progress of philosophical revolutions – but that is perhaps their only philosophical relevance. And some might even argue that they do not exhibit the degree of radicalism one may find oneself hearing in Kuhn's philosophical claims about either normal or revolutionary science.[12] For instance, Copernicus emerges from Kuhn's book *The Copernican Revolution* as a figure almost entirely bereft of revolutionary ambitions, and as someone who worked almost entirely within the Ptolemaic paradigm.[13]

We take our evaluation of Kuhn's essential identity to be confirmed by the kind of remarks which he made towards the end of his life, such as the following:

the new approach that has so fundamentally altered the received image of science was historical in nature, but none of those who produced it was in the first instance a historian. Rather they were philosophers, mostly professionals, plus a few amateurs, the latter usually trained in science. Though most of my career has been devoted to history of science, I began as a theoretical physicist with a strong avocational interest in philosophy and almost none in history. Philosophical goals prompted my move to history; it's to philosophy that I've gone back in the last ten or fifteen years; and it's as a philosopher that I speak this afternoon . . . (*RSS*, 106)

Given what I . . . call the historical perspective, one can [in fact] reach many of the central conclusions we drew with scarcely a glance at the historical record itself . . . [I]t is taking us [a long time] to realize that, with that perspective achieved, many of the most central conclusions we drew from the historical record can be derived instead from first principles. Approaching them in that way reduces their apparent contingency, making them harder to dismiss as a product of muckraking investigation by those hostile to science. (*RSS*, 111)

What is striking is first the degree of self-identification as a philosopher (of someone who began life as a physicist and then trained as a historian of science); and second the surprising and indeed trumpeted willingness to dispense with the density of actual historical examples in favour of the abstraction of 'first principles'. To say it again, Kuhn is a philosopher above all.

How Popper and Feyerabend go wrong

Popper and Feyerabend want to prescribe for science, preferring to prescribe *an attitude for individual scientists*, rather than a method. Both want to prescribe that scientists be adventurous, though each in their different ways. Popper wants scientists to be adventurously speculative in their constructive moments, and relentlessly critical otherwise. Feyerabend wants scientists to stop thinking they have got the sole truth on things, and to be much more anti-authoritarian in their practice. Popper and Feyerabend see the idea of *science as normal science* as prescribing a different attitude, that scientists should be obediently conventional, going along with the crowd. Both, however, can only make that criticism by exaggerating Kuhn's arguments about authoritarianism in the natural sciences.

Popper and Feyerabend think that the institution of science is in need of help from the philosopher (an idea perpetuated in the work of people like Fuller and other participants in the science studies movement), and both were primarily concerned with the place of science in a free society. This is not a concern Kuhn exhibits in his historical and philosophical work *at all*, being more preoccupied with descriptive issues than with prescriptive ones. This is not to say that his work is utterly devoid of 'prescriptive potential', only that this would be neither specific nor anything original, in many ways advising nothing other than continuing to do science – indignant at Feyerabend's accusation that he is ambiguous between description and prescription, Kuhn replies that they are to be read as 'both at once':

> If I have a theory about how and why science works, it must necessarily have implications for the way in which scientists should behave if their enterprise is to flourish. The structure of my argument is simple and, I think, unexceptionable: scientists behave in the following ways; those modes of behaviour have (here theory enters) the following essential functions; in the absence of an alternate mode *that would serve similar* functions, scientists should behave essentially as they do if their concern is to improve scientific knowledge.[14]

It is, we think, misleading to read Kuhn as *recommending* normal science. His most fundamental and contentious claim – though if it is heard right, it comes to sound like an utterly anodyne 'claim' – in that respect is that normal science *is what science is*. Meaning that: *there would be no science at all if there was not nearly only normal science*; that *revolutionary* science *cannot happen at all without a vast background of normal science*; and that *even in revolutionary times, most of science (and most of its upshots) are anyway partially like normal science and mostly unlike (say) philosophy*. The practice of normal science is deeply entrenched in the practice of the natural sciences, and is perpetuated by their organization. If normal science *is* what science is, then what value is there in recommending scientists to do this? It would be absurd, normally, for them to do anything else.

But this shows that Popper at least has misunderstood what normal science is, that he imagines that it is intended to highlight only one element in science, that 'puzzle-solving' picks out one kind of scientific activity, rather than advancing a characterization of (virtually) all kinds of scientific activity:

> The normal scientist, as Kuhn describes him, is a person one ought to feel sorry for. The 'normal' scientist, in my view, has been taught badly . . . He has been taught in a dogmatic spirit: he is a victim of

indoctrination. He has learned a technique which can be applied
without asking the reason why . . . He is, as Kuhn puts it, content to
solve puzzles.[15]

But the 'normal scientist' was never exemplified for Kuhn by
scientific inadequates, but by leading figures: Popper is hardly
entitled to feel sorry for Copernicus and Planck, or condemn them
as badly schooled.

If we concede anything to Popper's and Feyerabend's criticisms
we run the risk of confirming the view of Kuhn as indeed a conservative (which he may, for all we know, really be when it comes
to the matters that bother Popper and Feyerabend about the wider
society). However, we are saying his philosophy of science is not
usefully described as conservative; though it is hard to be heard
clearly, if at all, when one is saying something that does not fit on
a conventional spectrum. It happens as a matter of course in contemporary social sciences and humanities that thinkers are catego-
rized in terms of optimism and pessimism (which is usually much
the same as conservative/radical) about the possibility of generating improving social change. Kuhn, in these terms, would look like
a conservative, since, if you read into him the kinds of prescriptions
Popper and Feyerabend want to extrapolate from normal science,
he will appear to discourage people from thinking about the possibility of bringing about fundamental change in the nature of
science. Are we guilty of projecting a conservative Kuhn; or is Kuhn
actually guilty of the conservatism we 'attribute' to him? Neither!

Let us say that we see no reason to attribute to Kuhn any conclusion about the possibility or otherwise of improving democratic
control over science, or indeed about intensifying the corporate
control of business over its work. Concerns with the place of science
in society motivate disputes over these matters, and it is just not a
question arising within Kuhn's project, and no evidentially unsup-
ported inferences about these matters should be made.[16] The
description of science as normal science seems to us a depiction that
would not be altered by either the democratization or the commer-
cialization of science. 'Puzzle-solving' could be accommodated
within different forms of organization, as it was during the centuries when the natural sciences built up to their present position:
that scientists were once self-funding scholars, are now tenured
employees, and might increasingly become corporately sponsored
researchers does not figure in Kuhn's sketch, perhaps because it was
irrelevant to it?[17]

The real mistake is to think that engagement in normal science is a manifestation of an individual's attitude, one of mindless compliance with prescriptions, as though scientists were merely obedient with respect to prevailing requirements of scientific activity when they could as easily disregard them. But this picture is not anywhere implicit in Kuhn. The idea of the scientist as only the conventional follower of orthodoxy makes it sound as if the only thing inhibiting scientists from breaking with the orthodoxy is their timidity, their unwillingness to think for themselves – a pretty insulting view really. It makes it sound as if nothing could be easier than to think up an alternative way of doing things in any area of science, that nothing could be easier to come up with than ways of putting an existing theory on the rack (Popper) or of generating different but equivalently impressive alternatives that could match an existing paradigm (Feyerabend, and perhaps Pickering[18]). But this shows absolutely no sign of appreciating the extent to which Kuhn emphasizes how very hard it is to do anything (important) in science, and indeed, how little control any individual scientist has over the possibility of inducing fundamental novelty. The contrast between the normal and the revolutionary scientist is not, for Kuhn, as it is for Popper and Feyerabend, a characterological one, nor is it any difference in professional competence.

So, we need to make it quite plain that the idea that most scientists are not seeking fundamental change does not mean that they are somehow inherently conservative, intimidated by their training and professional experience from ever possibly thinking that their discipline might be usefully changed. We assume that what Kuhn does says does not conflict in any respect with the idea that every individual scientist might like to think of themselves as being eventually ranked with Newton, Darwin and Einstein, absolutely eager to make a complete breakthrough if only they could, that they might be permanently on the lookout for an opportunity to make their name in a really big way. It is not to be assumed that it is their personal conservatism that stops scientists from searching for fundamental novelty. Kuhn's point surely is that if scientists do dedicate themselves to changing fundamentals, if they were to decide that that was what they were going to do, then they would normally be making only an empty gesture, adopting an undertaking that would make no material difference to their scientific work. It would be like one of us deciding that life in modern society is pretty unadventurous, and that we had therefore better have an adventure, deciding to dedicate ourselves to seeking hidden treasure. This would be an entirely idle gesture without a more concrete idea of what treasures there might be, and of where to start looking for them.

To keep the argument concise, we'll try to round it out with two further points:

(a)   that fundamental work is dependent on timeliness; and
(b)   that the best way to make a fundamental contribution is to pursue normal science.

The latter is crucial and is the (achieved) burden, if read carefully, of chapters 5 to 9 of SSR (and the historical studies). Let us here recapitulate, and expand on the morals. If there really was a Holy Grail, we might set out on a quest for it; but there is no opportunity to do this, because we are not convinced there ever was any such thing, and have absolutely no idea where to start looking. And, in the same way, the situation in a mature science is that opportunities to make fundamental changes are relatively rare, and require a combination of (often fortuitous) circumstances – it is not plainly obvious where there might be room for a fundamental contribution, what sort of thing might make such a contribution, and so on, so it is not as if a scientist can identify the appropriate circumstances. Planck thought he was preserving the integrity of classical physics, not coining an idea that could eventually shake it – he just did not see the possibility of treating the discontinuity as resident in the phenomena.

'Did not see' might be 'could not see': recognizing that you might reject the 'continuum assumption', the assumption that matter was continuous, that particles were embedded in the aether (that you might seriously propose that this could be done in the context of nineteenth-century classical physics) could be a virtual impossibility. Not because scientists have a kind of sentimental attachment to classical physics, or are hidebound as a result of their career investment in that approach (though some of them might be). But just because it would be by no means easy to see how that idea could really cash out, or how it could be as much as a serious idea. What would it mean to talk about 'quantum discontinuity' in 1899? It is clearly hard for people even now, after a century of quantum physics, to get their head around the idea that nature is discontinuous and statistical, and one might therefore get a (correct) sense that Planck would have real difficulties in thinking that thought.

Now for point (b): we've been saying that a point which Popper and Feyerabend both miss is that the distinction between normal and revolutionary science considered as actual practice is not a sharp differentiation. For example, it is most important not to imagine that the distinction is supposed to be one which is reflected in the general attitudes of individual scientists, as if some were 'insurrectionist' and others were 'conservative' characters, some agitating for change, and others fighting against it – there are times, assuredly, in which scientists might be in these positions,[19] but it is not because one crowd is generally willing to knuckle under to orthodoxy and the others are not. As Kuhn tried to make plain, the scientific meat of a revolution is the estimation of the robustness and prospects of the paradigms, and there is often, taken from a properly historical point of view, little to choose contemporaneously between the two sides. Those who are reluctant to change may be no less sincere than those who will eventually win the accolade of history as the progressives, and it may be that those who resist change can give, argumentatively, as good as they get, have good arguments against their critics, see big flaws in the would-be rising paradigm. Remember, too, that absolutely nobody is proposing that they should go right back to the beginning, throwing out everything achieved under the old paradigm; the revolutionaries are proposing to go on from the paradigm in place, that is, to address problems arising in and from that paradigm, as well as to ingest parts of it in their new paradigm. Popularizers of science don't tell us much about the times when the resisters have been right and the would-be new paradigm turned out to be useless.

The difference between revolutionary and normal science is a difference in the state of the science rather than in the inclinations of individuals. There is some difference between the activity of scientists in normal and revolutionary science (they are, in the latter case, somewhat at a loss perhaps, and are casting about for ideas, and are spending time on propaganda, arguing fundamentals in ways that they otherwise might not) – however, this does not cash out as a difference between someone doing normal science and someone doing something quite otherwise, a kind of 'revolutionary science' imagined on the 'model' of (say) 'communist science' or 'new age science'. The revolutionary struggle is somewhat more like a propaganda war in which the two sides try to promote their respective causes and run each other out of town, but the scientific work that goes on within those periods and in the respective camps is not much different from the normal kind, in respect of puzzle-solving etc. – the revolutionary scientist is, as in Kuhn's main case studies, not out to make a revolution, looking for a way to turn the paradigm upside down and mostly arguing philosophy or methodology, but someone who has been trying to sort out a problem within the paradigm, has run out of ways of possible solutions within the paradigm, and is trying out new ways of solving the problem.

The 'advice' that the best way to achieve fundamental novelty is to do normal science isn't really advice at all; it is again more a matter of saying 'there isn't anything else to do than normal science.' Thus it is unclear how scientists might do otherwise than 'normal science', what kind of activities Popper and Feyerabend have in mind in addition to those Kuhn identifies. And this is one crucial dimension in which the real substance of Popper's and Feyerabend's difference with Kuhn is quite obscure. In so far as there could be substantial disagreement among them, it would seem to come down to an empirical question, though not one easy to answer without much research. Is Kuhn's list of activities that scientists engage in incomplete? Are there things scientists do that are not on it, and if so, how many of them are there? Second, if there are such activities, are they the sort that are definitely not 'normal science', that is, do not involve working on and within a paradigm, that do not constitute a puzzle-solving practice? Finally, if there are such activities, and they are not puzzle-solving, then how significant are they as a proportion of things scientists do – a big proportion or, relative to Kuhn's list, infrequent or even inconsequential?

Kuhn's key objection to Popper is with respect to the role of criticism. He accuses Popper of having an unrealistic picture of scientific practice. In the land of Popper's continuous revolution, science would not at all be science-as-we-find-it. At best, it would be like the pre-paradigmatic situation prior to the emergence of natural sciences, when everybody disputes fundamentals. Even worse, Popper's idea of criticism would make science more like the philosophy seminar than even a pre-scientific stage. Taking Popper in his barest terms, this would seem an appropriate objection. A more charitable treatment of Popper would be that he does not mean what he seems to say, that he does not and cannot in reality hold a picture of science as an endless and unrelenting round of critical argumentation. It can be argued that, in fact, both Popper's and Feyerabend's arguments presuppose Kuhn's picture of normal science (neglecting to notice that they do so).

Kuhn's objection is that Popper exaggerates the importance of criticism to science, and that it is (as anyone can see) just a matter of fact that working scientists do not constantly spend their time striving after revolutionary reputations, attempting to shoot theories down the instant they are put up. But is that really the way Popper thinks? One might suggest that each side is exaggerating the extent of their possible disagreement on this point because they have been drawn towards discussing the critical disposition of scientists at the level of individual motivation. Both sides, we suggest, are well aware that criticism is first and foremost an institutionalized feature of science, and that there is – in many ways – no real difference between individuals being critical, on the one hand, and going about their routine puzzle-solving work on the other.

Criticism is not manifested only by the slanging matches that might occur during revolutions, but is, or ought to be, crucially apparent in 'routine' scientific work. Popper does not, any more than Kuhn, really need to suppose that scientists are in endless, perpetual and fundamental dispute. Plainly they are not. But that does not mean, either, that they are not engaged in something worth calling criticism.

Perhaps some of the seeming divergence between them might then be explained in the following way.

Thesis   That Kuhn's focus is fairly heavily on scientists involved in investigations into phenomena ('finding out', as Hacking puts it), while Popper is ostensibly preoccupied with an image of scientists as testers of theories.

Argument   Kuhn's conception of normal science makes it the paragon of cumulative scientific work that contributes to the progress of our knowledge by the depth and intensity of its enquiry into nature.

A main part of normal science is the investigation of particular phenomena, with the normal scientist applying rather than contriving fundamental theory, relying on the accepted paradigm to deliver a result rather than striving to expose the theory at its most vulnerable points. Kuhn argues that though there is no testing of the fundamental theory against the results of the scientist's investigations, there is testing of the individual scientist's work against the standards of the prevailing theory. Normal science is puzzle-solving because the paradigm generates puzzles, which should be soluble in its terms by the competent investigator.[20] But it is, of course, a routine aspect of routine scientific work that it is presented to and criticized by colleagues – and, of course, it is an aspect of scientists' puzzle-solving efforts that these are often themselves essays in criticism of prior puzzle-solving efforts – the 'puzzle-solving' motivation including the urge to exceed the ingenuity of other solvers.

Thesis   That both Popper and Kuhn are tending to lapse into treating 'criticism' as a matter of individual motivation, as involving the difference between 'critical' and 'conforming' individuals rather than as an institutionalized feature of science.

Argument   The implication of the preceding argument is that 'criticism' is an institutionalized feature of the natural sciences, one which is embedded in the now utterly taken-for-granted forms of scientific activity, in the presentation, publication and circulation of scientific work, a level of intense scrutiny by one's colleagues that does not obtain in any other area of life. The extent to which Kuhn presents scientific training as 'dogmatic' is to suggest that it has the objective not of producing compliant and dogmatic individuals, but of producing competent practitioners, up to speed with current work, and capable of forging new directions of work for themselves and of significantly extending the reach of the paradigm, if nothing more than that. The natural sciences are not motivated to rehash past controversies,[21] and it is the business of scientific training to acquaint beginners with the received body of work. 'Critical ability' does not represent a skill acquired in addition to learning how to do the maths, design experiments, apply the theory and so on.

Thesis   That Popper leaves out of his story, in a way that Kuhn does not, the amount of work required to develop a paradigm to the point at which telling criticism is possible.

Argument   Kuhn stresses that a paradigm, on its first appearance, is very underdeveloped and its value lies in important part in its fertility as a source of further work that will flesh it out and capitalize on the possibilities it offers for raising levels of precision. Normal science provides the paradigm with its content, and does not fit a simple contrast between either seeking for refutation or seeking for confirmation. Spelling out and giving application to the paradigm is not a matter of seeking supportive evidence for it (let alone conclusive confirmation of it) as much as of mining its potential. Spelling out the paradigm will contribute to the potential for refutation, both by improving the precision of its claims, and by explicating its most unlikely implications and generating anomalies – thus exposing it to stronger risks that the results will not turn out as expected.

The notion of 'normal science' will subsume those individuals whom Popper would despise, the ones who adopt a cautious, bandwagon-riding approach, but it hardly includes only those, or intimates that they are more than a minority. It is clear that those who are practising normal science are often stretching both themselves and the boundaries of their science. It is not only those who create scientific revolutions who deliver notable scientific achievements. The alternative simply is not between the great revolutionary scientists and hack workers, and the overwhelming majority of scientists do – or so Kuhn thinks – stand somewhere between these extremes.

Finally there is the issue of 'anomalies' or potential falsifiers. Popper's notion of the 'verisimilitude' of theories can be seen as very much akin to Kuhn's idea of notable scientific achievements, at least in this respect: both recognize that there will be pros and cons for any theory. Each assumes that though a given framework fits nature, it does so only to a degree, and that, somehow, inevitably, subsequent science will reveal respects in which it does not. However, as Kuhn argues, and Popper effectively assents,[22] scientists do not give up a theory just because it is (in at least certain respects) 'falsified'. Scientists give up one theory only in order to take on another. The scientific revolutionary has to match and surpass the achievements of the prior paradigm, and it is only under some conditions (unspecifiable in advance) that the potential of some anomaly to the prevailing paradigm to provide the basis for a surpassing alternative conception will be discerned and developed. There may be conservatism among some scientists, and there may be an unwillingness because of vested interests in the status quo to entertain suggestions that the prevailing paradigm is fundamentally flawed, but this does not explain why anomalies only occasionally become the basis for the refutation and replacement of the paradigm with which they will not fit.

In sum, then, with respect to the issue of the critical nature of science, if Popper and Kuhn are read charitably, not as expounding extreme positions, one holding that science is unRealistically critical and the other that it is unRealistically dogmatic and authoritarian, there need be little disagreement of substance between them. Popper cannot help presuming what Kuhn emphasizes, that revolutionary science turns over a paradigm by putting another in its place. Kuhn highlights what Popper at least downplays, namely, the difficulty in coming up with any serious scientific conjecture, that is, something that could viably act as a new paradigm.

The thing that Feyerabend neglects about normal science as Kuhn portrays it, and which renders his prescription for paradigm proliferation vacuous, is the categorical nature of scientists' procedures under paradigms. They do not treat the paradigm as a provisional installation, a point of view that can be opted for or declined according to one's preferences. The paradigm is not a 'point of view' on phenomena, which is to be offered in a spirit of interpretation, towards which alternatives might be entertained. The paradigm's policies are treated as definitive of matters of fact, and their outputs categorically stated. But this undermines Feyerabend's objective, which is to recommend that scientists should take a more distanced attitude towards their creations, express them in tentative terms, an inclination that would be acceptable if it was legitimate to cultivate alternative paradigms. Kuhn's account indicates that Feyerabend's proposals would require that the practice of science be changed very drastically, quite fundamentally, to legitimize the proliferation of paradigms, and involve, as such, the tacit admission that such proliferation currently has no real place in science, and that, therefore, Kuhn's 'normal science' does depict what science is actually like.

But isn't this why Feyerabend's proposal is necessary? Well, to think that it could be at all plausible from Kuhn's point of view really is to overlook what Kuhn has laid out with respect to normal science and the pursuit of paradigms: the creation of paradigms is not a for-its-own-sake affair, and actual paradigms emerge from the scientific daily round and are only retrospectively identified as such, and it would be a very fundamental change in science to value paradigms as ends, and to value therefore a plurality of them. The very identity of something as a scientific discipline is often tied up with unity-around-the-paradigm, so it is quite unclear as to how different paradigms could actually exist, how they would retain any sense of being related to each other. The conditions for the formation of a new paradigm are tied in very specific ways to their predecessor, a fact that Kuhn formulated with his emphasis on the technical bite that they must have. That one can formulate one paradigm with sufficient technical bite to mount a challenge is rare enough, that one should generate a plurality of them . . . As

Feyerabend envisages them, the plurality of scientific ideas would lack just that feature, the technical bite, that, in Kuhn's account, marks out a paradigm.

Further, paradigms are not produced *de novo*, they are in important part constituted out of the prior paradigm, and thus a new paradigm cannot accede autonomy to alternatives, for it will want not merely to conceive things in its own terms, but, as part of that, to reconceive the prior paradigm's achievements. What would be the value to working scientists of attempts to develop new paradigms just to ensure that there was a plurality: how great an effort would it be to engender a new paradigm significantly, if not entirely, different from that already in place? If it was merely to be an optional alternative to the other, then the effort would have to be just as successful as that already in place, and therefore just as problematic to achieve as the initial paradigm was. Further, what is the point of developing an alternative paradigm just as good as that in place? If going to the trouble to develop a brand new paradigm, why not produce one that is better than the alternatives? What is the point of having a (presumably mostly notationally equivalent) mere alternative to a paradigm – why not contrive a replacement for it?

We do not offer this in ideological defence of normal science, but as a suggestion that Feyerabend had not really thought through his proposal, that he had not seen the extent to which he depends on the idea of normal science to promote his critique, but also the extent to which, in trying to articulate it, he has failed to address the features that Kuhn identifies as essential. In consequence, he has no answer to the question: how would one leverage change away from the normal science which currently comprises the natural sciences, towards the kind of thing that (Feyerabend thinks) he would prefer to see.

Perhaps a further important element in Kuhn's difference from Popper and Feyerabend is the extent to which Kuhn treats science as primarily an investigative activity, while Popper and Feyerabend really treat it as a predominantly theoretical one. Though Kuhn never puts the point in so many words, he sees that the value of the examples are in their fertility as an inspiration to researches into specific kinds of knowledge. It would be wrong to think that what scientists were doing was 'exploring their paradigm' *rather than* investigating particular kinds of phenomena: finding out about particular kinds of phenomena is exactly the same as exploring the paradigm. Normal science involves, then, for example, finding out about the chemical reactions involving proteins, the nature of regularly repeated energy emissions from stellar bodies, the effects of chemicals on the brain, the conditions under which metals fatigue, etc. The interest in 'normal science' can lie at least partially, if not entirely, in finding out more about the nature of a particular phenomenon.

Kuhn is trying to tease one away from a ubiquitous and deep-set intellectualism in thinking about science. Kuhn's account of 'normal science' can, controversially, be seen as a novel model of what it is for a collectivity of persons to be engaged in rational inquiry. Of course, many forms of rational inquiry are not science – but arguably, normal science can be read as a model for rational inquiry in general. A model emphasizing that much must be taken for granted for anything to be disagreed upon, that a focus on advancing research requires a disciplined exploration of an agreed agenda, etc. This would be Kuhn as philosopher *par excellence*, for this would be a mode of argument going well beyond sociology or history. But it would still be philosophy in the sense of being a (non-intellectualist) approach to scientific activity.

Consider again – try to *see* – Kuhn as (a) revolutionary in the philosophy of science. The concept of normal science is not a conservative conception: it is very threatening, it is radical, and it was largely hitherto unrecognized as vital, foundational, for the study of science, for studies of science which do not take the finished discovery or theory as the only relevant standpoint from which to tell a tale of scientific change, and which do not focus on the rare unexpected innovation made by an isolated or eccentric or heroic figure as the be all and end all of such change. It is threatening, because it says that little if any of science is like the Popperians said it was. And because even scientific revolutions in their socio-historical concreteness emerge out of the same processes, not literally out of odd isolated eccentrics with bold and brave new ideas. This thought is in a sense stronger than Kuhn's important and well-known claim that 'Anomaly appears only against the background provided by the paradigm. By ensuring that the paradigm will not be too easily surrendered, resistance guarantees that scientists will not be lightly distracted and that the anomalies that lead to paradigm change will penetrate existing knowledge to the core . . .'; 'So long as the tools a paradigm supplies continue to prove capable of solving the problems it defines, science moves fastest and penetrates most deeply through confident employment of those tools' (*SSR*, 65, 76).

The distinction between normal and extraordinary science, so severely criticized by Toulmin and others, is in a way of less import once one has understood that the actual practice of science need not be that different in order for its results to be extraordinary. To anticipate, we want to claim that the point of talk of incommensurability can be separated from the distinction between ordinary and extraordinary science, at least in so far as that distinction is thought of as a matter of the kind of practice engaged in at the time, rather than as a matter of conceptual transformations fully visible only to the sensitive historian. What one sees when one thinks through the normal versus revolutionary distinction with care, is that the kind of thing that is going on in science (not particularly in its individual practitioners) at these different 'moments' is such that:

(a)   revolutionary science, which in its day-to-day substance is mostly normal science anyway, could not emerge except from normal science; and

(b)   unless one understands that, conceptually, there are the normal and the revolutionary moments in science's development, in the 'structure' of science, then one will fail to understand science at all. That is, accounts of science that left no room for the kind of stability of problems and data found normally, and for the kind of conceptual innovation found occasionally, would not really be accounts of science at all.[23] They might successfully account for some bit of scientific activity; but that would dissatisfy their producers. They mean to be accounting for science – once they understand Kuhn's approach and criticisms adequately, they will (or should) give up even their partial account, for it does not achieve what they wanted to achieve in the first place.

In a certain sense, one might go so far as to say that the real criticism directed at Kuhn by Popper et al. was that Kuhn is not sufficiently the philosopher of scientific revolution. Popper basically says to Kuhn: The problem with you is your notion of normal science, or at least your notion that normal science is basically OK. I could stomach your talk of 'scientific revolutions' if only you were prepared to admit that (little) scientific revolutions are happening all the time, or at least ought to be.

Popper plays not Roosevelt but rather Trotsky, to Kuhn's (as imagined by Popper) Stalin. Popper preaches not, as one might expect, no revolution and instead enlightened liberal efforts at rational change – no, one might say he preaches permanent, 'global' revolution. He looks for a culture in which scientists are continually prepared to radicalize their enterprise from the ground up, in which there are no shibboleths, only the ever-present prospect of drastic change and the jettisoning of basic assumptions.

Popper's primary accusation – that people are, for Kuhn, stuck in a framework, that he is an irrationalist in thinking that free unframed thought is impossible – is of course the very opposite of the accusation that others make most central, that Kuhn is an irrationalist in thinking that drastic change is too easy, in preaching the gospel of scientific revolution. (It is perhaps worth noting right here that Kuhn emphasizes the internality (that is, resulting principally from factors internal to the science in question), even 'necessity', of scientific revolutions. And after all, even Gestalt switches are not unconstrained. Quite the contrary: one can see a duck or a rabbit, but little or nothing else. As intimated above, Popper has an inadequate idea of what being a revolutionary would involve. It is not a matter of just saying, 'Let's be revolutionary!' It is a matter of being willing, if push really comes to shove (as it rarely does), to countenance more fundamental change/revolution than Popper is willing to countenance.

One might risk describing Kuhn as providing the basis for an anti-Foundationalist epistemology, with at its centre a reconceptualization of the nature of rational inquiry focused around the nature of normal science. Perhaps seeing Kuhn as doing this makes him look less extreme than *either* those who accuse Kuhn of an 'anything goes' mentality *or* those who accuse him of being a 'frameworkist' would have it . . .

Kuhn's revolution in the philosophy of science is, we would suggest therefore, rather widely misunderstood by those not closely familiar with it, and perhaps unaware of its threat, force, or promise. Barry Barnes is arguably right in this, at *least* so far as 'the social sciences' are concerned: Kuhn's most radical successful contribution to the philosophy of science is probably not his particular rendition of 'scientific revolution', nor exactly his (much-misunderstood) notion of 'paradigm-shift' – it is in his role specifically as the philosopher of normal science that Kuhn's most authentically revolutionary contribution is to be found. It was in his insistence on the mechanisms through which science is and has to be normally *non*-revolutionary that Kuhn struck a blow to the heart of the Popperian enterprise, and depicted a 'new' vast realm of science, one which conformed to the standard (post-)Positivist image of science only at the cost of rendering the latter more or less trivial, because Kuhn explained how the cumulativeness of normal science was relative to a paradigm (a disciplinary matrix), whereas for Logical Empiricists and their successors (including now many recent/contemporary philosophers who take themselves to be Realists, and oppose the anti-Realism of the Carnap etc. crew), the cumulativeness of knowledge was absolute.

In sum, 'fans' (such as Barnes) and 'foes' (such as Popper) alike appreciate (albeit often only dimly and foggily) that, once one gets beyond the banalities of the loosest talk of 'new paradigms' consequent on catastrophic/dramatic-sounding scientific revolutions, Kuhn's suggestive philosophy of science is what most people are interested in. And that philosophy of science is founded perhaps less on concrete historical accounts of ruptures (and still less on actually established sociological investigations) than on a novel and abstract account of the normal, skilful, mundane procedures of problem-solving, puzzle-solving (communities of) scientists. The latter may sound unexciting, but, as we have stressed, its consequences are arguably just as revolutionary as all the talk of 'scientific revolutions'. Kuhn's account of normal science 'parallels' his account of scientific revolutions: he introduces, through the two put together, truly an account of science . . . for the first time ever.

We hold, then, that a fatal flaw in most criticisms of Kuhn on scientific change is that they try to criticize one of the 'aspects' or 'moments' of science as described by Kuhn, while surreptitiously drawing, at the same time or elsewhere, on that very moment. Thus, as already mentioned, Popper attacks the legitimation of normal science as science. But he also requires the kind of stability there is in ordinary (normal) science in order for there to be enough agreement for falsification to actually work – to work stably in a community. Furthermore, Popper fails ultimately to account even for the genuinely revolutionary moments in science, because he fails to make room for the kind of significant conceptual change (SSR, 98) for which the ground is prepared only by normal science (SSR, 88–95).

Carnap and other Logical Empiricists ((post-)Positivists) seem to understand the existence and importance of normal science, but not only does their narrow reading fail to leave any space for revolutions, they don't really leave sufficient space for normal science. They don't leave space, that is, for exploring the paradigm, for puzzles (SSR, 98–101).

Feyerabend is a limiting case. He seems to overcome the different limitations of vision of Carnap and of Popper alike, but at the cost of obviously yielding an activity that will bear scant resemblance to anything we now call science, that will look more like (say) Modern Art. Feyerabend really has no time for normal science. In him alone does one find an ideal type of someone picking one of Kuhn's two aspects and focusing all his praise on that, all his ire on the other. And we have wondered whether Feyerabend would be prepared to countenance the real abolition of normal science.

Mostly, the important thing to see is how virtually all would-be methodologies of science, even arguably the Feyerabendian 'anti-methodology', actually presuppose Kuhn's picture of science. This is why Kuhn says, in his subtle way, the extraordinarily strong critical things he says of the received view in the philosophy of science. Namely that, if we take them at their word, rather than seeing how they collapse into his view, the Positivists and all their successors and inheritors (this includes the Falsificationists, the Logical Empiricists, and arguably also Feyerabend – as an anarchized post-Popperian – and all who retain any 'formalism' in their methodology and epistemology, any notion of the scientific method as expressible in philosophical/logical formulae[24]) fail to give us a genuine view of science *at all* (cf. *SSR*, 80, 98–102). Taken at face value, even late Carnap, and sophisticated Popper, and Kuhn's 'friend' Feyerabend . . . fail to generate accounts of science at all.

PHILOSOPHY OF SOCIAL SCIENCE

One of the reasons why Kuhn's ideas were so influential on the social sciences and humanities was because these were already much interested in and *influenced by* the philosophy of science. Kuhn's book must have seemed like yet another, and welcome, contribution to the efforts that would-be social scientists had long been making, namely, to turn themselves into proper scientists, under the guidance of the philosophy of science. Thus, there was abundant enthusiasm in some circles for Popper's advice: sociologists and their like should become scientists by ensuring that they developed proper theories, ones that could be refuted by evidence. Kuhn's message must have seemed reassuring in this way: it strongly reinforced the idea that the social sciences *could really be sciences* – in a manner easier to achieve even than adhering to the Popperian form.

Certainly, Steve Fuller is intensely critical of Kuhn on the grounds that many people have taken Kuhn's account of the natural sciences as a model to adopt and follow in the human sciences and social studies. That cannot be, however, a basis on which to criticize *Kuhn*, for it was not *his* advice that these people were misguidedly following.

Consider, then, the situation that social scientists have long found themselves in. They can recognize this situation in Kuhn's description of the early stages of the natural sciences. The early natural sciences have no paradigms, they have schools, ones which are endlessly debating fundamentals, with each inquirer trying to make a completely fresh start, and to rebuild the whole discipline anew and from scratch. But this is just what contemporary social sciences are like. Kuhn's study shows how the natural sciences have been transformed, how someone – a Newton, a Galileo perhaps – has come along and *succeeded* in entirely reshaping their area of inquiry, turning it into a unified discipline. There must, then, be hope for the social sciences: their present condition is just like that of many scientific pursuits before they developed unity-around-a-paradigm. Therefore, there is nothing to say that social studies cannot and will not turn into paradigm-ruled pursuits, and seemingly every reason to expect that they will. Is it, though, just a matter of sitting around and waiting for one's Newton or Galileo to turn up, or is it a matter of making a more deliberate effort to *implement* the promise that can be found in Kuhn? In other words, has Kuhn shown us how to turn the social 'sciences' into the social sciences?

Let us begin by setting out briefly what we argue for in this section. Our argument, in simple terms, is this: that the social sciences make the mistake of supposing that the application of the philosophy/methodology of science in their own area is a key to scientific progress. It is the old and seemingly endlessly attractive idea of modelling one's efforts on those of (other, more successful) sciences. But, Kuhn asks, is the success of the natural sciences because natural scientists know more about *truth* than anyone else, or just because they know more about plants, or animals, or soil, or stars or whatever their interest might be than others. 'It may, for example, be significant that economists argue less about whether their field is a science than do practitioners of some other fields of social science. Is that because economists know what science is? Or is it rather economics about which they agree?' (*SSR*, 160–1). How much do economists or sociologists – or physicists or geologists – really know about *science*? How much *does anyone* really know about

science in either the sense of knowing how science is actually done
and actually works, or in the sense of knowing what it is about the
things that scientists do that make them *scientific*? What social 'scientists' think they know about science is merely the *philosophy of
science*. The implication of what Kuhn says is therefore that, really,
social scientists typically know nothing or next to nothing about the
actual natural sciences.

So, social scientists are going about things in entirely the wrong
way in so far as *they are guided by the idea of attempting to turn what
they do into sciences*. This, kind of argument is very strongly put by
Kuhn with respect to the social sciences (so called) in his argument
about the futility of the methodological imperative to quantify (cf.
'The functions of measurement in modern physical science', *ET*,
179–221, esp. 221). The fact that the most successful (not by any
means all, we add) of the natural sciences are highly quantitative
has convinced many that the very essence of science is measurement. This conviction has certainly been influential on the social
sciences, and the imperative of measurement – Go forth and quantify! – disfigures (especially) American sociology even today. Kuhn
argues that the attempt to build a science through quantification is
entirely futile, and that the history of science shows this. The history
of science shows that quantification often arises out of a rich and
detailed *qualitative* knowledge – or at least, aesthetically / practically
productive and impressive conceptualization (think of Copernicus,
for instance) – of the phenomena in question. Creating methods of
measurement *independently* of one's understanding of the phenomena one is attempting to investigate just is an unproductive exercise. One requires above all an exemplary scientific *achievement*.
Not simply a quantificational form to one's hodge-podges or
hand-wavings.

Kuhn's own approach is really telling the social scientists that one
cannot move from a pre-paradigmatic state to a paradigmatic one
by design, that none of the paradigm-based pursuits have emerged
in this way. Thus we can perhaps polemically summarize Kuhn's
implication in this way: that the attempt to make (say) sociology
scientific by modelling it on the natural sciences is just about *the
most unscientific* way of proceeding imaginable. Effectively, the
attempt to set out the ground plan of the discipline, and to build on
that, is precisely to perpetuate the condition of 'the schools', and to
try – imagining that one is following the way of the natural sciences
– to do what no natural science has, to our knowledge, ever done.
(Of course, we don't mean that one natural science can't borrow

ideas and examples from other natural sciences, and in that sense, follow the example of another science, for they do, of course, routinely do that. We're talking rather, about the attempt to build a science *from scratch*: one has a title for the discipline – sociology, say – and one then tries to figure out in advance what that science will consist of.)

There is a standard, yet, we think, *wrong* image of Kuhn which, in outline, is as follows. The way to scientificity is simply the *establishment of* the dominance of a *paradigm* (disciplinary matrix) in one's subject. One needs to professionalize one's discipline around the central focus of the doctrines of one of its schools, and that's all. The pro-Kuhn camp takes the message of this to be 'We just need a paradigm – if we enforce agreement on a paradigm, then we can be a proper science. There's nothing more to being a proper science than that.' The anti-Kuhn camp takes the message to be 'Kuhn legitimates Relativism and mob rule, because any group of people can "get" a paradigm in that sense – so the genuine sciences will not be able to distinguish themselves from pseudo-sciences if Kuhn's ideas are accepted.' Kuhn is saying that it will be in their own good time that disciplines acquire a paradigm, if ever they do. They cannot be frog-marched into a paradigmatic state.

We are not trying to put into Kuhn's mouth the argument that the social sciences cannot become *bona fide* virtually-indistinguish-able-from-the-natural-sciences sciences. The line of thought we have just intimated carries no implications either way about how things will turn out, but certainly does suggest that, however they do turn out, there is nothing *predestined* about the development of a paradigm (or lack of same). Kuhn's whole approach gives us no reason to assume anything other than that the emergence from pre-paradigmatic status is, so far as anyone can judge, a wholly *contingent* matter.

Mary Midgeley provides a useful example. Now, someone who had read Kuhn, and who as a consequence of such reading had come up with the popular thought that what one had to do was indeed to try to get the methods of one's own school to triumph as '*the* scientific method' in (say) psychiatry, would then presumably wage an essentially political campaign to demonstrate and enforce this, replete with accounts of how the beliefs of their school alone could enable psychiatry to understand itself and refine its methods and get lots of government money and be respectable among the medical community, etc., etc. The question is, would this be likely to be a productive thing to do? Would it be likely to hasten the advent of a truly scientific psychiatry?

Midgeley addresses these thoughts by means of querying in particular whether Materialist schools of psychiatric thought are pursuing their efforts to scientifize their discipline in a way that ultimately makes much sense:

> [T]he reduction of mind to body is now seen as a major factor in determining diagnoses and methods of treatment. As two concerned practitioners in this field have put it,
>
>> 'Despite the ambiguity and complexity of psychiatry, it is striking that many students begin its study with the appearance of having solved its greatest mysteries. *They declare themselves champions of the mind or defenders of the brain* . . . The unfortunate result is that many of them become partisans – and needless casualties – in denominational conflicts that have gone on for generations and that they scarcely understand.' (emphasis in original)
>
> As the authors point out, this metaphysical issue cannot be ignored. . . . It is
>
>> 'more than a question of taste whether we think about schizophrenia as a clinical syndrome . . . as a set of maladaptive behaviours, a cluster of bad habits that must be unlearned, or as an 'alternative life style', the understandable response of a sensitive person to an insane family or culture.
>>
>> Each of these proposals makes different assumptions about the phenomenal world and its disorders, and each has different consequences for psychiatric practice and research . . . The result of ignoring the fundamental differences between perspectives is not to diminish sectarianism but, in the end, to encourage it.'
>
> It seems reasonable to suggest that they would best be seen as viewpoints belonging to investigators encamped round the mountain of mental trouble. *Yet the temptation to choose one and to take sides is extremely strong for a profession that feels the 'scientific' imperative compelling it to choose only one approach.*[25]

Thus Midgeley is claiming that the idea that there ought to be one of the existing schools triumphing in order to commence a glorious new age of normal science within a discipline which as yet lacks it is part of the *problem*, not part of the solution. The effort to make one's discipline scientific may well encourage sectarianism, rather than diminishing it, and to no productive end.

Are Feyerabend's criticisms valid?

Bearing in mind our earlier discussion of 'paradigms' and 'normal science', let us focus in on one of Kuhn's rare direct remarks on the social sciences in this context:

> In parts of biology – the study of heredity, for example – the first universally received paradigms are [really quite] recent; and it remains an open question what parts of social science have yet acquired such paradigms at all. History suggests that the road to a firm research consensus is extraordinarily arduous. (*SSR*, 15)

The use of the word 'yet' might imply a teleological vision; and likewise the phrase that Kuhn uses elsewhere, 'pre-paradigmatic'. But just because certain disciplines have become . . . disciplines, have become sciences, surely cannot imply that all will. For example, here is one possibility: that the social sciences will eventually come to appear to most of us as astrology appears to most of us now – as a pathetic attempt to ape science, failing because it never achieved a genuine tradition of research, a genuine actionable set of problems and puzzles.[26]

But the immediate point is this. We hope it will not be taken as a mere vacuity if we remark that Kuhn describes for us the structure of normal science and of scientific revolutions – *in those disciplines which do in fact fit the description*. In disciplines which 'find themselves' with paradigms. In disciplines where the expression 'pre-paradigmatic' is not necessarily misleading . . . because the discipline did as a matter of fact come to have a paradigm, did come to be unified around an exemplar(s), etc. Kuhn lays down no advice or prognostication for disciplines without a paradigm. There is *nowhere* in Kuhn – not in the quote given above, nor anywhere else – a *claim* that one can confidently predict that in a discipline with schools, the eventual victory of one school can be confidently predicted. We do not mean simply: the victory of one particular named school. No, we mean the victory of any school, ever. Kuhn's claim concerning the emergence of paradigms is purely a *retrospective* claim. *He is talking about the structure of the emergence of those disciplines that have become sciences*, not suggesting *how non-scientists can become scientists*. Not providing a manual for the creation of new sciences. He is, at least by implication and omission, pretty clear that there can never be a guarantee that a discipline without a paradigm will acquire one, and thus no sense in which it can be obvious and perspicuous that (for instance) the social sciences are well conceptualized as on the road to normal science.

For the victory of a school, the construction of certain types of
institutions is perhaps necessary. And certainly afterwards. But this
does not imply that it is a good idea to construct such institutions
at any particular time. Nor that the construction of these will ever
be *enough*. One needs to have sets of agreed-upon exemplars,
common methods/ways of acting, and an absence of ongoing foundational disputes. There are strict limits to the extent to which any
of these can be imposed on others in the discipline unwilling to
accept their imposition. One can try to suppress foundational
disputes – for example, through hegemony in a professional association or in educational institutes in a discipline – but this is liable
to be to some extent self-defeating, especially in any climate valuing
academic freedom, etc.

Let us sum up. The attempt to *force* a victory of one school, the
forced establishment of some dominant exemplars, and thus the
imposition of a disciplinary matrix for the first time will normally
result, not in a surer road to science, but in a surer continuation of
the reign of 'schools'! *Kuhn ought neither to be praised nor buried for
having apparently given 'pre-paradigmatic' sciences a road or a menu
towards normal science.* Because, appearances to the contrary, he
simply did not do so. Careful reading indicates that he did not even
attempt to provide such a road or recipe.

For Kuhn aims to record primarily not what scientists say they
do, nor what others say they do, nor what they think they should
do, nor what others think they should do, but *what they actually do*.
Thus it is off-target for Feyerabend to refer to Kuhn as having the
'belief' that scientists 'should not waste . . . time looking for alternatives' to a working paradigm; and for him to claim that Kuhn is
saying that history can tell you the way science *should* be run rather
drastically misses Kuhn's point.[27]

Feyerabend and Kuhn again

But let us note also two positive moments of Feyerabend on Kuhn:

(a)   Feyerabend, much more than the orthodox Popperians,
      found Kuhn's concrete accounts of scientific *revolutions*
      impressive and highly suggestive – suggestive of the extent
      to which such events are 'non-rational', and even also have

an aspect to them worth calling something like 'incommensurability'; and

(b)   rather more surprisingly, there can also be found in Feyerabend, if one searches it out, a different strand – namely, a limited but real defence or explication of Kuhn's *normal* science.

In particular, in relation to (b), Feyerabend helps us to account for something which can be elusive, something which was perhaps still partly missing in the previous section, and which Barnes does not supply in the slightest: that is, why Kuhn's remarks on normal science are couched so thoroughly at the level of generality that they are, why they are not in fact concretely exemplified *prior* to (or at least after) the generalizations, why therefore it can rightly look as though what Kuhn is supplying us with is at 'most' the outline of a 'philosophical sociology'.

Here is Feyerabend, providing us with the materials to fill in that gap, to complete perhaps our account of (Kuhn on) normal science:

> [In the analysis of science] historical research and not rationalist declarations must now determine the nature of the entities used, their relations and their employment in the face of problems and . . . a *general* theory of science *must make room* for these specific parameters. It must leave specific questions unanswered and it must refrain from premature and research-independent attempts to make concepts 'precise'. Kuhn's account perfectly agrees with these desiderata. His paradigms are 'obscure and opaque' not because he has failed in his analysis but because the articulation changes from case to case. The relation between theories and paradigms remains unresolved because each research tradition resolves it in its own way, in accordance with the cosmological, normative, empirical elements it contains.

Thus there can be a general account of normal and/or extraordinary science only at a very high level of abstraction and generality. Feyerabend makes it clear that Kuhn's paradigm concept is (thus) not pitched at such an *unacceptably* high level – it could not be pitched otherwise if it is to be pitched at all (if one is to do more than *only* tell specific historical stories – and if one were only doing the latter, one would have entirely given up the philosophy of science). If one attempts to make it more methodologically detailed and concrete, one is back (as Feyerabend sees it) to the fantasies of late-Popperianism and Lakatosianism, or to outright Positivism. If one wants a philosophy of science at all, at *least* for the purposes of uprooting other (wrong-headed) philosophies of science, one will *inevitably* then produce something more like a philosophical sociology than like a concrete first-order sociology.

This highly useful sympathetic (as opposed to warning/highly negative) Feyerabendian interpretation of Kuhn *even on normal science* is to be found actually at a handful of important points in Feyerabend's work; for instance, compare the following:

> Understanding a period of science (according to Kuhn) is similar to understanding a stylistic period in the history of the arts. There is an obvious unity, but it cannot be summarized in a few simple rules and the rules that guide it must be found by detailed historical studies (the philosophical background is explained by Wittgenstein . . .). The *general* notion of such a unity, or 'paradigm', will therefore be poor and it will state a problem rather than finding a solution: the problem of filling an elastic but ill-defined framework with an ever-changing historical content. It will also be imprecise. Unlike the sections of a theoretical tradition which all share basic concepts, the sections of historical traditions are connected only by vague similarities. Philosophers interested in general accounts and yet demanding precision and lack of ambiguity . . . are therefore on the wrong track; there are no general and precise statements about paradigms.[28]

Do you need a paradigm to catch a paradigm?

There is one point further to the above arguments worth making here, in this chapter, and right away. What were folks expecting when they expected that Kuhn's *Structure* should show them a way forward for the human sciences, for their own social discipline, along the sure road of science? These people, these sociologists and political scientists and psychologists and anthropological linguists, and perhaps also literary and art critics and historians and students of religion, were presumably expecting that Kuhn was pointing them towards something beyond his own practice. But if they had looked at that practice rather harder for rather longer, they might not have been so expectant that it would provide them with something radically new, with the outline of how to get hold of a paradigm *for their* use.

For one is struck, if one notes the listing of disciplines just made, that *Kuhn himself* borrows something from each of them. That is, one could well bear in mind his use of metaphors from religious studies and politics, his analogies to art history, his use of methods from textual criticism, his borrowings from Gestalt psychology and from Sapir-Whorf, not to mention his straight history (and sociology). So, Kuhn's own practice draws far more on the practices of the human sciences and the humanities than does that of any other major philosopher of science. Kuhn is to some considerable extent *himself* a human scientist. As intimated above, Kuhn's own approach can be modelled on that of a natural scientist (and thus his account of scientific revolutions self-applied) only at considerable cost, or with considerable care. Kuhn is doing philosophy, and is doing it after the fashion of interpretive human science far more than most philosophers of science (some of whom attempt to draw more on Logic, or 'Cognitive Science', or to build their own '(philosophical) *theories*' of science, for example).

He attempts to 'capture' how science works through employing the concept of 'paradigm', etc. And one hopes that what Kuhn is finding is not wholly of his own invention. In other words, one hopes that 'paradigms' are actually already there in the social practices of the communities etc. which Kuhn is talking about. In other words, Kuhn encourages one to look at the sciences and to see if one finds paradigms there. Paradigms are already objects of Kuhn's philosophical sociology. Exemplars and disciplinary matrices are both constitutive of the order of scientists' practices in a normal science situation. And again, they pre-exist Kuhn's description of them, they are not just artefacts of Kuhn's writing (unless Kuhn is quite wrong – in which case no one would need worry about his lessons for the social sciences).

Now, what is it that human scientists more generally do? Is it not something very similar? For example, a sociologist looking at any set of practices, whatever they may be, or an anthropologist looking at an 'alien' society: are they not in search of the beliefs, methods of acting, tacitly agreed norms, etc., among the people they are looking at? Are they not in the business of describing these?

We are suggesting the following: that, rather than taking inspiration from the *content* of Kuhn's descriptive history of the natural sciences, rather than looking for a paradigm to guide and enforce limitations on their own practice *qua* social scientists (an enterprise liable to be useless or even counterproductive, as argued earlier in this section), social scientists could profitably look instead to Kuhn's version of their practice *as he himself employs it*. Social scientists could usefully think through the sense in which their own practice is *already* the searching out and describing of paradigms and their ilk. Do they *need* a separate paradigm of their own to do that properly? Might it not rather hinder their task? Take our sociologist, looking (say) at some set of religious practices and beliefs. Isn't Peter Winch right in saying that 'the sociologist of religion will be confronted with an answer to the question: Do these two acts belong to the same kind of activity?; and this answer is given according to criteria which are not taken from sociology, but from religion itself.' Why? Because:

> The concepts and criteria according to which the sociologist judges that, in two situations, the same thing has happened, or the same action performed, must be understood *in relation to the rules governing sociological investigation*. But here we run against a difficulty: for whereas in the case of the natural scientist we have only to deal with one set of rules, namely those governing the scientist's investigation itself, here *what the sociologist is studying*, as well as his study of it, is a human activity and is therefore carried on according to rules. And it is these rules, rather than those which govern the sociologist's investigation, which specify what is to count as doing 'the same kind of thing' in relation to that kind of activity.[29]

Similarly, the appropriate criteria for deciding whether two people are engaged in the same kind of activity – are both testing a particular hypothesis or not, say – belong to that activity – say, the specific science in question – itself. At times of paradigm shift, there may suddenly be divergent decisions among scientists on such issues, such decisions.

Let us sum up the above. Kuhn's 'fans' say: social science needs a paradigm (that doesn't exist yet). We say: social science is about, among other things, finding paradigms *that exist already*, in social settings. (And what one surely needs to do in relation to such paradigms is to describe them, to be responsive to them as they already exist, not to impose an alien theory on to them.) Students of society should not think of the practice of finding paradigms and exploring their nature descriptively as radically new. Kuhn did what they have *already* been doing, to a considerable extent (and vice versa, now). And what they do looks very different from what natural scientists do – for the latter, unlike the former, do not have as their business *anything like* the description of paradigms. If they 'explore' paradigms, it is in the utterly different sense indicated in earlier chapters – through theorization, experimentation, etc. And the point we have made is that one has no particular reason to think that the human sciences will profit from aping such methods. For they need essentially to explore something like paradigms *which already exist*, rather than creating a new one. They need to effect a description of a set of human practices, a set of practices which do not necessarily need to have a paradigm or a set of categories or suchlike imposed upon them in order to become interpretable, for they already *embody* such a set of categories. They may even be argued to be correctly describable in principle in a sense unavailable to natural scientific inquiries, because there is no such thing as describing the non-human world in a way that the latter prefers. Whereas perhaps the human world truly can be 'cut at its joints' (!) – by a description that respects it and (simply) gets it right, gets it in terms accurately reflecting participants' self-understandings and ordered activities. Paradigms and their ilk already exist – *in* social action (for instance, of scientists).

The 'Kuhnian' apologists for social science may, in their desperation for respectability, be overlooking this possibility, the possibility that careful description of social action, of paradigms etc., may as Mary Midgeley among others suggests, already be possible ('even' in disciplines – the social studies etc. – without a paradigm), provided one doesn't get sidetracked into fighting for dominance of a would-be social *science*. And let us be quite clear about this: the wish of Kuhn's cruder fans or appliers to ape natural science by means of 'getting' a paradigm, finding their own Newton, is deeply ironic. These people, who think that Kuhn has proved a kind of Relativism to be true, are still – at the very same time – wanting to have the kudos of being recognized as scientists, by means of having a paradigm to unify them and the paraphernalia of professionalism to maintain and enforce the unification. But this shows that they are still vulnerable to the attractions of Positivism or of 'Scientific Realism'. If they really had the confidence of their Relativist convictions, they wouldn't care about how the natural sciences conducted themselves, they wouldn't try to ape science as described by Kuhn or whoever – they would boldly strike out in their own direction, they would rest content with self-generated criteria for how, if at all, to distinguish between good and bad ways of structuring their discipline, good and bad work within their discipline. It is in fact a sign of deep disciplinary insecurity that one calls upon a philosopher of science to supposedly legitimate one's own discipline as being 'just as good' as the natural sciences. It makes no difference whether one hopes to do that by pulling one's own discipline up to their level, or pulling their discipline down to one's own level . . .

The alternative of course is for 'the social sciences' to regard themselves as truly *sui generis,* as not needing to look to methodological aspects of the sciences with paradigms in order to validate themselves. This can only be done if one more or less accepts the current state of one's own discipline, ducks out of endless methodological debate (except in so far as it is necessary to puncture the aspirations of those, for example, who have been criticized above) and *gets on with* doing what good work can be effectively done within that discipline. For example, perhaps, in sociology, good ethnographies and descriptions of very diverse social practices. Including (but only as one case among many) of scientific practices.

The community of practitioners has to actually be impressed by – has to actually pretty much universally recognize – a scientific achievement, and take it for a paradigm. If that doesn't happen, then too bad – *you don't have a science.*

Kuhn does not dwell very much on borderline cases of (mature) sciences. One extrapolates to these cases at one's peril – our interpretation of the true (and small, if you like) impact of Kuhn on the philosophy of the social sciences is far less perilous a way to go. And we have been clear that Kuhn was not in the business of laying down norms for how to *get* paradigms, or saying that 'getting' a paradigm is always possible.

The clinching piece of evidence for our interpretation, with which we end this chapter, is from Kuhn's longest, if still fairly brief, consideration of an example from the social sciences in *SSR.* We have already mentioned the moment; let us now dwell upon it:

> Why should the [scientific] enterprise . . . move steadily ahead in ways that, say, art, political theory or philosophy does not? Why is progress a perquisite reserved almost exclusively for the activities that we call science?
>
> Notice immediately that part of the question is entirely semantic. To a very great extent the term 'science' is reserved for fields that do progress in obvious ways. Nowhere does this show more clearly than in the recurrent debates about whether one or another of the contemporary social sciences is really a science. . . . Men argue that psychology, for example, is a science because it possesses such and such characteristics. Others counter that those characteristics are either unnecessary or not sufficient to make a field a science. Often great

energy is invested, great passion aroused, and the outsider is at a loss to know why. Can very much depend upon a *definition* of 'science'? Can a definition tell a man whether he is a scientist or not? If so, why do not natural scientists or artists worry about the definition of the term? Inevitably one suspects that the issue is more fundamental. Probably questions like the following are really being asked: Why does my field fail to move ahead in the way that, say, physics does? What changes in technique or method or ideology would enable it to do so? These are not, however, questions that could respond to an agreement on definition. Furthermore, if precedent from the natural sciences serves, they will cease to be a source of concern not when a definition is found, but when the groups that now doubt their own status achieve consensus about their past and present achievements. *It may, for example, be significant that economists argue less about whether their field is a science than do practitioners of some other fields of social science. Is that because economists know what science is? Or is it rather economics about which they agree?* (*SSR*, 160–1; emphasis added)

Here is the crucial difference between *naturally acquiring* something like a paradigm, and merely *striving deliberately* to get the *trappings* of one. Kuhn's 'followers' and his 'antagonists' alike confuse the latter with the former.

4

Incommensurability 1: Relativism about Truth and Meaning

Kuhn and the philosophers

Can we get Kuhn off the charge of 'semantic relativism'? (Do we need to?) Kuhn has been attacked not only by the leading philosophers of science, but also by the leading American philosophers of his time, and the most important of these – such as W. V. O. Quine, Donald Davidson and Hilary Putnam – have accused him of a semantic relativism that is self-defeating, involving a fundamental, and almost elementary mistake. Does Kuhn say that there are irreducible discrepancies in meanings between different scientific theories, between different cultures, *between languages*, so that systems of meaning are reciprocally closed? Put in the way the issue actually came to be discussed: is it possible to translate without loss between one 'system of meaning' and another? In defending Kuhn against these critics we do not suggest that it is *all* down to their misconceptions of what Kuhn says, for the determination of what Kuhn means by what he says on the issue at stake here – incommensurability – is not easy. Nonetheless, we will deny that Kuhn's argument has the fatal flaw that these critics allege.

The limits of understanding

Kuhn says that pre- and post-revolutionary scientists 'live in different worlds' (*SSR*, 121) and, unlike Feyerabend and Imre Lakatos, insists that they can live in *only* one world. The scientist is always 'in the grip' of one paradigm, that which is dominant at the time. The scientist, therefore, cannot understand what came before the revolution? But if the scientist *cannot* understand more than one paradigm, then how can the historian or philosopher hope to? But Kuhn, as either historian or philosopher, then seems in this impossible situation: he tells us that we cannot understand previous science, and yet writes case studies to explain previous science to us.

Donald Davidson's attack on 'conceptual relativism' in 'On the very idea of a conceptual scheme' is heavily sarcastic about Kuhn in this respect: Kuhn wants us to imagine radically different – incommensurable – conceptual schemes in science and is also aware that we can only ever be in one of those.[1] How then can Kuhn possibly understand those different from his own, let alone understand them in a way which enables him to explain them to the rest of us? So it is a mystery as to how, if two things in the same field are incommensurable, we can understand both well enough to see that that is what they are. Surely, if incommensurability is a strictly semantic matter, then we cannot think of both paradigms, hold them in mind, and then 'notice' the incommensurability. If they are incommensurable, then *that's that*: any paradigm other than our own must be effectively unrecognizable to us. (It would seem a result, then, that we cannot even assert the thesis of incommensurability itself.)

Incommensurability: what does/can it actually mean?

Literally, incommensurability means the impossibility or unavailability of a common system of measure. The idea originally comes from mathematics, from the shock the Pythagoreans got long ago from realizing that there must be numbers that are not expressible as fractions, that are not 'rational'. This in fact follows almost directly from a thorough understanding of Pythagoras's law concerning right triangles. If one has a right isosceles triangle, and if each of the shorter sides is one unit in length, then the hypotenuse, by Pythagoras's law, must be the square root of (one squared plus one squared equate) two units in length. So there is an object in the real world which can be shown to be the square root of two units in length. But it can easily be mathematically shown that the square root of two is not a 'rational' number; for if we suppose it to be a 'rational' number, then a contradiction follows.[2]

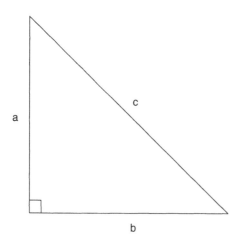

A ruler, however minute its divisions, could not measure side c absolutely accurately if it could measure sides a and b absolutely accurately. The measurements in question can be done successfully in ordinal terms, and indeed to any arbitrarily required degree of accuracy; but there is always the possibility of demanding that the measurements be done more accurately, at which point the 'incommensurability' will again be evident.

Therefore there are objects one of whose sides has a length that is in an important sense not comparable to the lengths of the others. The latter can be expressed as fractions, the former simply cannot. There is in other words no common system of measure, in a certain important and quite intuitive sense, to (for instance) the sides of some right triangles. The 'irrational' numbers are 'incommensurable' with the 'rational' numbers.

This idea is taken up by Kuhn and Feyerabend and applied to science, where it would appear that there may be measuring schemata – schemata to 'measure the world' with – which are not intertranslatable without loss of data or meaning.[3] However, one should note that there is a sense in which the mathematician can nevertheless *describe* the 'incommensurable' numbers perfectly well – otherwise one couldn't understand the incommensurability just described in (we hope) an accessible way! And there are ways in which the numbers – the lengths – in question can be perfectly effectively compared: for example, we can say that the square root of two is greater than 1.4, less than 1.5, etc., etc. Similarly, Kuhn and co. must be able to describe to us the incommensurability they are concerned with, on pain of there being no idea that they are putting forward here. The question will be whether this necessity of their being able to express the incommensurability means that it cannot cut very deep at all. For example, must any lack of intertranslatability perhaps be absolutely marginal/minute, as seemingly it is in the maths case? [4]

Slightly metaphorically – and this is of the essence of incommensurability according to Kuhn[5] (Feyerabend argued at times for a similar version) – the idea he seemingly *wants* is that there may be ways of doing science, *weltanschaungen, paradigms*, which are logically discontinuous: they don't simply contradict each other, since they are different in ways which deny them the resources respectively to make statements which flatly confront each other. If they could do that they would be commensurable, but they can't. And it is at the least an open question whether the mathematical analogy alone will get Kuhn all the way to this.

An example of 'incommensurability'

One of Kuhn's key examples of incommensurability, as we have already noted, is the transition from Newtonian to Einsteinian physics. One cannot fault Kuhn for bravery here. He takes on his opponents on one of their strongest turfs. For it is virtually a verity, and a widely accepted one, that Newtonian mechanics is true as a special limiting case of Einsteinian Relativity (*SSR*, 97–8). Popularizing scientists and philosophers alike have thus found in this famous case a prototypical example of the traditional image of scientific progress: the prior view remains true, but only within a restricted domain of application, while scientific knowledge grows seamlessly to cover novel cases efficaciously, cases for which the prior view was false or inapplicable.

Here is the heart of Kuhn's account of how, even though it appears that Newton's Laws can be derived from Einstein's relativistic dynamics, this appearance is actually delusive:

> [T]he physical referents of [the] Einsteinian concepts *are by no means identical with those of the Newtonian concepts that bear the same name. (Newtonian mass is conserved; Einsteinian is convertible with energy.* Only at low relative velocities may the two be measured in the same way, and *even then they must not be conceived to be the same.*) Unless we change the definitions of the variables . . . the statements we have derived [from Einsteinian Relativity, in an attempt to show Newtonian mechanics to be a special case of the latter] are not Newtonian. If we do change [the definitions], we cannot properly be said to have derived Newton's Laws, at least not in any sense of 'derive' now generally recognized. [The] argument [of those who would

derive Newton from Einstein] has, of course, explained why
Newton's Laws ever seemed to work. In doing so it has justified, say,
an automobile driver in acting as though he lives in a Newtonian universe. An argument of the same type is used to justify teaching earth-
centered astronomy to surveyors. But the argument has still not done
what it purported to do. It has not, that is, shown Newton's Laws to
be a limiting case of Einstein's. For in the passage to the limit it is not
only the forms of the laws that have changed. *Simultaneously we have
had to alter the fundamental structural elements of which the universe to
which they apply is composed.* (SSR, 102; emphasis added)

The last point is crucial. Kuhn argues that Newton and Einstein take
the universe to be populated by different fundamental entities.
There is no way, he is saying, for one to intertranslate between the
two without obliterating this vital fact. If one reads (for instance)
the word 'mass' differently from how it was read under the
Newtonian paradigm, one cannot literally be claiming to derive
Newton's laws. In scientific revolutions, *the furniture of the universe
changes.*[6] And there is no neutral further language in which to
describe this furniture (SSR, 114–15, 125–7). There is only, normally,
the language of the victors, which tends either to make its predecessor look stupid or bizarre *qua* scientist, *or* to make it look as if it
was always really trying to be its successor. Thus Newton is usually
made to look 'proto-Einsteinian'. Alternatively, he could be treated
as for example Aristotle sometimes is – as a great philosopher, but
as someone whose physics (whose dynamics, especially) is just
'crazy', and hardly worth calling 'science' at all.

To put it another way: in so far as Einstein superseded Newton,
this could actually only be so in tandem with the belief that Newton
was wrong – Einstein didn't show that Newton was a special case of
his own, new, view. (Of course, we must be very careful if we say
this; we must bear in mind that for Kuhn there is never simple *refutation*, never a crucial experiment that one must be convinced by; the
observations of Mercury during the eclipse which are sometimes
taken as a crucial experiment that proved Einstein's theory correct
depend on so many further assumptions etc., and were in any case
only achieved well after Einstein's theory had largely won the day –
on no 'evidence' at all, one might say. One might then say that for
Kuhn there is only refutation as such when there's *no contest* any
more, that is, in a context of pure justification alone. The refutation
is only clear when one theory has been decisively accepted at the
cost of another – but by that time, the other is hardly *understood* at
all; and, if understood at all, probably only by means of the efforts
of historians and philosophers of science. The predecessor is made

to look stupid; or bizarre and crazy in their views such as to be not really a scientist at all; or (often best of all) rather *like oneself*. All three can be Whiggish strategies, but the third is the most effective if it can be pulled off, for then it becomes impossible to read predecessors as themselves. They read rather as if they've been trying to be you all along, *as if they've read you*, but ill-understood you.)

Can't one still say that 'There just *isn't* mass like Newton said there was; his classification system was wrong; mass *à la* Newton doesn't refer'? We can respond to the question briefly (and effectively) as follows: *No*. One can't unmisleadingly say that, because classification systems in themselves are not true or false. Only things that are said using classification systems can be true or false; but then there has to be some 'system' within which their truth or falsity is assessed. So, we can if we wish say of Newtonian statements about mass that they are in some instances false because they fail to connect up with reality at all: but then we'll *already* be tending to misread them; we'll already be reading them from the perspective of the Einsteinian paradigm or some such.

We can't simply say ' "mass" *à la* Newton doesn't refer', except as an elliptical way of already *making* the same Einsteinian critique of Newton. ('Reference', for Kuhn, does not make an isolated connection between words and world, but refers (!) to what a Relativist would call a purely '*intra*paradigm' phenomenon.)[7] To say (for instance) ' "mass" *à la* Newton doesn't refer' is only to show that we have already accepted Einstein's system. That system is/must be authoritative for us, now, *qua* physicists or scientifically informed laypersons. The Einsteinian could truly and non-question-beggingly say, '*We* just don't use the term "mass" in the way that Newtonians did; we have *good reason* to believe that the theories one gets out of using it that way will be less good than those we can get out of talking *à la* Einstein instead.' Thus, it's not exactly that one can't use the term 'mass' in the full-blown old way (except in certain restricted and simplifying/pedagogical contexts), it's that we just *don't. We don't take that usage seriously.* Except if we are interested in understanding the *history* of the discipline. And again, what natural scientist *qua* natural scientist should have to be much interested in doing that?

Understanding as translation

Let us go deeper into the question of understanding science, understanding past scientific ideas. As we have said, argument about

incommensurability gets turned into argument about translation –
because the two hugely influential philosophers, Quine and
Davidson, both think that understanding is pretty much the same
as translation. Clearly, in one respect this is true, but it is surely a
mistake to tie understanding and translation (or 'interpretation' as
Davidson calls it) entirely together as Quine/Davidson do, attempting *to understand the first in terms of the second.*[8]

The idea that understanding and translation are wedded might
seem persuasive, for surely to understand someone else's ideas is a
matter of being able to put them into words that we ourselves can
understand? Thus, if someone speaks to us in a foreign tongue, we
cannot understand what they say, for we cannot understand their
words. To understand them we must put their words into our
English ones. 'Translation', however, does not apply, in the thought
of someone like Quine, only to transactions between a foreign and
our native tongue, but also to cases that are within ostensibly the
same language. We have to assume that the words of the other
speaker are used with the same meaning that our words have, that
they *do* translate into our own meanings.

Kuhn is thus being drawn into one of the central disagreements
in modern American philosophy, reflected in the following quotation from Hacking, over whether there can be any unique, determinate translation between one language and another:

> There are three philosophical fantasies that we could label 'too
> much', 'too little', and 'just right'. The just-righters . . . claim there is
> just one right system of translation between any pair of languages
> . . . Philosophical debates currently [in 1975; and still to quite a large
> extent in 2002] focus on the claim that there is *too much* free play
> between languages to determine any uniquely best system of trans-
> lation. The most famous exponent of this is Quine, who calls it the
> indeterminacy of translation. Imagine that we observed all there is
> to observe about speakers of an alien language. We know every occa-
> sion on which any sentence was, is, or will be uttered, and we know
> all the observable precedents and consequences of each such occa-
> sion. We even know how speakers of that language are disposed to
> talk in situations they never in fact experience. In short, we know
> infinitely more than any radical translator [an anthropologist encoun-
> tering a quite alien language] ever could know. Even if we knew all
> that, there are, claims Quine, indefinitely many mutually incompat-
> ible systems of translation that would square with the data.
>
> Quine urges that there is *too much* possibility for translation.
> The opposed doctrine maintains that there is *too little*. Two human
> languages could be so disparate that no system of translation is

possible. This is in the spirit of Feyerabend's doctrine of incommensurability.[9]

Hence, the problem of incommensurability tended to be conceived both by the critics and by Kuhn himself to be one of the possibility of translation. Kuhn is willing to say that his doctrine of incommensurability does signify the impossibility of thoroughgoing translation between two different scientific schemes. Cast in those terms, and understood in the context of the continuing debate over translation, Kuhn seems to confront the following difficulty: in denying that there can be translation between two schemes, either

(a)   he is wrong in that there can, there *must* be translation between such schemes, and his whole (philosophical) project collapses (since it is, for Kuhn to maintain his 'incommensurability thesis', essential to make the case that there is no such possibility); or

(b)   he is right, there is no translation between them, in which case, his whole (historical) project collapses (since he, no more than anyone else, can translate a defunct paradigm into our contemporary one, and hence *cannot possibly* explain their worldview to us).

Davidson insists (supported by specific philosophical reasons than need not matter to us here) that we should not call something a language that we could not translate into other languages. Therefore, the idea of separate 'conceptual schemes' between which there would be no translation does not hold up. If conceptual schemes are things between which we cannot translate, this means there can be no such things as distinct conceptual schemes. Thus Davidson believes that Kuhn clearly should abandon incommensurability (a), but that, given that Kuhn plainly does not want to do this, that he (Kuhn) ends up in position (b), without the ability to say the specific things about episodes in the history of science which apparently prompted his serious turn to philosophy in the first place.

We believe that Davidson however does not allow Kuhn any philosophical space, that he unduly restricts the room for manoeuvre that there is on this issue – that, in effect, he is much too quick, and as a result begs the question against Kuhn.

Kuhn against Davidson

Let us begin our concrete response to Davidson with what might seem a relatively small point: Davidson too readily generalizes Kuhn's argument to 'language' from its designed application specifically to the case of the scientific scheme. Davidson may be right about the context of conceptual scheme talk vis-à-vis natural language, but this is really irrelevant to Kuhn's argument. Like Davidson, and unlike Sapir-Whorf, we think that English and French (and even Hopi) are not usefully regarded as constituting or directly yielding different conceptual schemes. But does that imply that theoretical systems of science etc. are not reasonably called conceptual schemes? No;[10] *philosophical work* would need to be done to make the implication go through. As we discussed in chapter 2 above, Kuhn, in his account of the Copernican revolution, identified some functions for conceptual schemes, such as those of saving on memory, facilitating the working out of implications and so on. Kuhn would not have needed to – for example – identify the Ptolemaic system as a conceptual scheme, nor to make arguments about the functions of such schemes, if he had regarded conceptual schemes as identical with natural languages. Part of Kuhn's specific point about 'scientific' conceptual schemes, which he later came to emphasize strongly,[11] is that their constituent terms are interdefined. Thus, Davidson's denial that different languages house different conceptual schemes does not directly contradict Kuhn's insistence that the Ptolemaic astronomical system was developed into a conceptual scheme.

Further, we have mentioned (p. 60 above) the ambiguity in Kuhn's initial formulation of incommensurability as something that might be understood as either thoroughly globalized or highly localized. Kuhn later unequivocally disambiguates his meaning: incommensurability is a *local* thing, even in the context of a scientific theory, something which affects only those areas of a science in which there are interdefined terms. Thus, the impossibilities of translation between conceptual schemes of the kind Kuhn has in mind are *partial* only.

What's more, Kuhn *never said* that it was impossible for scientists to understand their predecessors. Kuhn said (documenting his claim with historical research) that it was *possible* for them to do so, and that they *did* do this, even that there were *conditions* (incommensurability being one) *favouring* both likelihood of misunderstanding between rivals in a revolutionary period and of retro-spective misinterpretation of prior work. This is, for Kuhn, an unsurprising fact, a consequence of his argument that it *is* impossible to translate *entirely* between two conceptual schemes. One cannot express in the terms of one theory everything that can be expressed in the terms of the other. But this does not make understanding *both* paradigms impossible, only very tricky, requiring (imaginative and historical) effort, work.

Kuhn does not, therefore, accept that understanding and 'translation' are wedded in the way that Quine and (especially) Davidson think they are.[12] In understanding how Kuhn can insist both that translation is impossible, and that understanding of 'different conceptual schemes' is possible, we need to look again, and more suspiciously, at the idea of translation. Quine's and Davidson's idea of what translation is should not be thought of as being the same as what is involved in translating French speech into English as we ordinarily go about this. We have words in English such as 'totem' which are words that could not be translated into English. That is, there was no single word in English which meant what 'totem' meant in the languages that explorers encountered, so, of course, we have *added* the word to English. We could not, then, translate 'totem' into English, in the sense that we could not identify one word which would (did) match the word. Of course, we could get English speakers to understand the word 'totem', by doing a lot of explaining (in English) about the people who used the term, how it fitted into what they do and, thus, what they meant by it.[13]

Kuhn himself, of course, has been busy 'translating' historical science, sometimes written in now-defunct tongues, into English; but, of course, his 'translating' is not a matter of matching word for word the texts which express those ideas, but of giving us elaborate explanations of their meanings, which draw on much more than narrow linguistic understandings.

Are we now ourselves disagreeing with Kuhn and saying that he is wrong: translation is possible, he does it himself? Worse, are we impaling Kuhn on a deadly (and more or less Davidsonian) contradiction: he denies that it is possible to translate at the same time as he is doing exactly that? In fact, we are drawing attention to the fact that the claim that translation is impossible depends essentially on what one understands by 'translation'. Here is Kuhn:

> I [suggest] that the problems of translating a scientific text, whether into a foreign tongue or into a later version of the language in which

it was written, are far more like those of translating literature than has generally been supposed. In both cases the translator repeatedly encounters sentences that can be rendered in several alternative ways, none of which captures them completely.[14]

What Kuhn is denying is that the kind of translation that Quine and Davidson ostensibly seek is available between two 'conceptual schemes'. *In principle*, a translation would give, for each sentence in one language, a sentence in the other language that would match it in meaning, and one could do this for all the sentences in the two languages. But what if the sentences were incredibly long? What if the 'sentence' needed to translate a sentence from a scientific text was actually the length of a monograph, having to go into all the historical conditions and sensibilities etc. involved in the old conceptual scheme, so as to avoid missing its real 'meaning'?

Now Kuhn's objection has become clearer: it is that between two different scientific systems it may (sometimes) be impossible to find for *certain* terms in the one system *any* terms in the other that match them in meaning. It may just be that one system has, and the other recognizably lacks, a term (as with the totem example). It is obvious that modern physics has no term which corresponds to 'phlogiston'. But, any term or even any longer phrase which one might hope to get to correspond with phlogiston characterizes the world in a way which cross-categorizes with modern Physics's way – the two cannot be rendered compatible. Sometimes, it may be that *it looks as if there is an appropriate term*, as if the two systems *share* the term in question. This may be a delusion, and even though one can produce a sentence in system one which matches *word for word* a sentence one can construct in system two, it *still may* be impossible to claim these two sentences as translations for each other! Why?

We can give a first approximation to an answer in the following fashion. Because some of the words differ in meaning (in one sense of that word), as with the example we (following Kuhn) have used – 'mass' in Newton's scheme does not mean the same as 'mass' in Einstein – so, the word-for-word resemblance between matched sentences of the two schemes would be superficial only. Translation is meant to match two sentences that mean the same, and assertions in Newton's scheme about mass *just don't* mean the same as they do in Einstein's: in the sense that the two matched sentences would lack the same 'feel', the same *detailed set of interconnections*. The only way one could get that feel or atmosphere, that same set of detailed interconnections, would be through a large-scale exercise in history, philosophy and imagination – that is, through the kind of thing that Kuhn pursues in greatest depth in his case studies. And it would be odd indeed to say that the translation of (say) 'The Earth is at the centre of the Universe' is to be found simply in the totality of the argument of *The Copernican Revolution*. Better to say, not just that there is no *point for point* translation between some Ptolemaic and 'Copernican' statements, but that there is an important sense in which they cannot be intertranslated at all. Would one call something a translation into Russian of 'A bird half wakened in the lunar noon / Sang halfway through its little inborn tune' if it was the length of a book, in order to capture all the possible allusions of the lines in English, etc., etc.?

The impossibility of translation in the word-for-word, sentence-for-sentence sense follows from the nature of scientific systems, or conceptual schemes as Kuhn sometimes terms them. These schemes have internal interdependencies, and *some* of the vocabulary may be interdependent. Kuhn's point applies *to cases in which terms are inter-defined*. If that is so, if a term's meaning (its use, broadly considered) hinges partly but crucially on its relation to certain other terms, then to pluck the term from that context is to divest it of its meaning. One *cannot* just *transplant* that word to another system: attempting to do so will simply result in changing its meaning because one will be relating it to a different set of interdefining terms. Thus, *where* two scientific systems are closely related as rival paradigms there will be *points* at which no translation of particular sentences between them will be possible because the key terms cannot be isolated from the respective vocabularies of the two schemes.

Let us recapitulate what has been shown above. The failure of 'point by point' translation is important, but it is not the conclusion of Kuhn's argument. The point which Kuhn brings out particularly in his consideration of attempts to translate between phlogistonic and 'modern' chemistry is this: that not even long 'paraphrases'[15] or complicated multiple uses of symbols can be usefully said to 'capture', to *translate*, phlogistonic chemistry: 'We have not only to say that phlogiston sometimes referred to hydrogen and sometimes to absorption of oxygen, but we have to convey the whole ontology of phlogiston in order to make plausible why it was taken to be a single natural kind.'[16]

The work of attuning oneself *to a different sensibility* is the work of interpretation. It may in some cases, even if carried through to the maximum feasible degree, not result in what Kuhn wants to call a translation, in, we might, say, the full sense of that word. Because sometimes, Kuhn thinks, there is no such thing as a translation available between two modes of thought. They are not missing a common currency in the sense of some extractable sense, or 'sameness of extension'; these only apply in an intra-paradigmatic sense. There can be different modes of thought, styles of reasoning, 'grammars', whose difference can be 'detected' by the intelligent work of history, and by an 'aesthetic' sense.

Equally important to Kuhn's rebuttal of the accusations on translation is his assumption that *understanding comes before translation*. It is perhaps appealing to think of being unable to understand something said in a foreign language, and then being able to understand it because it is translated. It is important to understand, now, that it is a distinct peculiarity of Quine's to think that the same point applies to our own language too, that when someone speaks to us in words that seem exactly like our own English words, nonetheless we are involved in translating them into what we imagine are our own equivalents. Quine may have philosophical arguments to justify treating understanding that way, but we are not trying to resolve the dispute between him and Kuhn, but to point out that the two are at cross-purposes.

Kuhn treats the native and alien tongues cases as different, in accord with our more usual idea that we translate between two different tongues, not within the same one. Then the apparently persuasive link between understanding and translation collapses. We did not learn to understand our *native* tongue through translation. When we acquired our mother tongue we had no other tongue to translate it into. Thus, we learned our native tongue directly, we learned to understand it *from scratch*. Thus, understanding proceeds independently of translation in this case. In the case of translation between two tongues, think of it from the point of view of the translator, rather than from that of ourselves as people who understand what someone said in another language because someone puts it into our own for us. Someone who translates has to be able to understand *both* languages before they can translate. Someone who understands French cannot translate passages in French for us if they have no English. Thus, it is only someone who understands both languages who can *establish* that one language has a word 'totem' for which the other has no equivalent.

The historian of science is like the anthropological visitor to a remote tribe, having to start to learn their language from them, even from scratch. Someone who is a bilingual understands both languages, but that does not prevent there being parts of one language which cannot be translated into the other (the word 'television' might present such a problem for a language other than ours; much more serious difficulties, difficulties which cannot even in principle be solved merely by the addition of a new word into the other language, are presented by literary language).

Let us summarize Kuhn's conception, especially as laid out in his later works, of his views on incommensurability, as expressed through the concept of translation:

> *Interpretation*, a fundamentally hermeneutic concept, is suggested by Kuhn to be fundamental to understanding another language (or whatever) when one *does not already understand that language*. (Thus, importantly, Quinean 'radical translation' is better read as a kind of interpretation, according to Kuhn.)
> *Bilingualism* results, if one manages fully to acquire the new language.
> *Translation* is possible only when one can move between the two languages in a manner which enables one to 'capture' in the one language what is happening in the other. Slightly less vaguely, one must be able to follow and render the network of significances and associations that matter to *meaning*, in a 'full' sense of that word, before one can be said to be able to *translate*.

A parallel with Peter Winch

It is worth mentioning Peter Winch's arguments about some problems in understanding an alien society,[17] which involve very similar concerns and responses to Kuhn's, but also make explicit something that is only implicit in Kuhn. Kuhn had compared the historian of science's difficulties to those of an anthropologist who must come to terms with another society with very different ways than the anthropologist's own.

Winch, treating these kinds of difficulties as in a way representative of general problems in the social sciences, focused on a specific case, that of understanding a tribal society that engages in magical practices of the sort that can seem quite incomprehensible and bizarre to us. How can people possibly believe that the things that they do can work? Winch's effort, in their case, is very much the same as Kuhn's with Aristotle, Copernicus and Planck: to show that this question poses a false problem, that if their way of thinking is properly identified, if its terms are fully elaborated, then we will come to see that the fact they can and do engage in those practices is no more problematic for them than it is for us to be fully immersed in our own practices. The problem, Winch is arguing, arises not from the nature of the alien belief system itself, but *from our assumption* that our way of thinking is not only *the correct* one, but that it is utterly obviously so. *Unless* something prevented those people from doing so, they would naturally see the world the way we do – Winch, like Kuhn, wants to remind us that our way of viewing the world is by no means 'the natural one' (for there is no such thing as what one tends to want from that expression), and the impression that it is the natural and obvious one results from our familiarity with and immersion in it, not from our understanding of how we came by it in the first place or of the ways in which it resembles or differs from other ways that people *have also* found persuasive.

What does Winch make explicit that Kuhn does not? As we have presented the argument over translation there seems to have been an attempt (by Davidson) to force a choice between two *extreme* positions. *Either* we deal with different conceptual schemes, which are utterly closed to each other, so that the inhabitant of one cannot possibly understand another *or* it is impossible for there to be serious divergence between two systems because they must be fully translated one into the other. We have established that Kuhn does not fall into either of those positions, and that there is room to be taken up by him (and Winch) between them (or, if you prefer, orthogonally to them). This, of course, is by virtue of Kuhn's holding, *as he actually does*, that two scientific schemes are neither wholly closed against each other nor fully translatable one into the other. Winch's point is that attempting to understand an alien system is not just a process of 'translation' but is one *of learning*, a matter of changing ourselves and adding to our own conceptions – we do not understand an alien society by absorbing what it does into what we do, but come to accurately appreciate the difference between our ways and theirs, recognizing that we have been apt to treat such cases in the past with condescension, rather than understanding.

Thus Winch makes readily apparent – makes explicit – an intermediate 'position'/possibility between ordinary understanding on the one hand and sheer nonsense/nothingness on the other: namely, what it is to understand something worth calling different . . . something which can be understood only by an effort of imagination or analogy, something which resists one's categories. Something, for example, like a genuinely new philosophical idea. This 'intermediate position', providing one sees it, shows that, and how, Davidson begged the question in 'On the very idea of a conceptual scheme'; and how one can say the kind of thing that Kuhn and Winch want to say without committing oneself to a self-refuting 'semantic relativism'.

Relativism about truth

We've denied that Kuhn (and Winch) are 'semantic relativists'. Are we/they out of the woods? Not yet. There is a worse suspicion: that Kuhn, and Winch, are *relativists about truth*. The relativism seems to follow from incommensurability. Here is what incommensurability *means*: *in so far as* there is incommensurability between different scientific paradigms, *then so far* one is unable to translate (in the strict sense) what can be said in the words of one into what can be said in the words of the other. If you can say 'p' in the language of scheme 1, you can't say 'p' in scheme 2. In that case, of course, you can't say 'not p' in scheme 2 either. In other words, you can't get schemes 1 and 2 lined up so that they flatly contradict each other, so that one says 'p' and the other says 'not p'. Hence, you can't decide between them both by comparing them with the actual situation (whether or not p), since to describe the situation as p or not-p is to use the meanings of one scheme, when the legitimacy of using those meanings to describe the facts is just what is contested.

Does incommensurability not then automatically *imply* relativism about truth? If there is no *scheme-independent* way of telling whether 'p' is true, and that therefore that means that it is *only in terms of scheme 1* that it is possible to say that 'p' is true, then if it is true in 1 it cannot be said to be false. But doesn't that mean that what any scheme says is true *is true just by virtue of the scheme's saying it is true?* And wouldn't that mean that what every scheme says is true is true *because* the scheme itself says it? And isn't that just Relativism: every scheme is as justified as every other.

But didn't we, in our explaining of 'incommensurability', say, in the original mathematical example, that that we can't compare in terms of a common measure at all doesn't mean that we can't compare (at all). Didn't we also say that Kuhn does insist that scientific change involves comparison of paradigms, and what is a scientific revolution but a judgement that one paradigm is *better* than another.

Someone determined to extrapolate a relativist implication from incommensurability will not necessarily go away at this point. All very well, they may say, but the comparison is only between one paradigm and the other, but a comparison between one paradigm and another can have no purchase: there is no role for nature in this comparison. But isn't the polarity of contrasting paradigms with each other, on the one hand, and contrasting paradigms with nature, on the other, just the polarity that Kuhn explicitly rejects: *in real, historically existent science, comparison with nature is made by way of paradigms.* And those comparisons, as carried out by science, *are decisive* (though they may not be simple, uncomplicated and patently conclusive). What else are the Kuhnian case studies except an attempt to help the reader understand *why* that simplistic picture of two theories proposing *p* and its direct contradiction *not-p* each being compared against the independently established fact of *whether or not p* just does not occur.

So Kuhn is *not* a Relativist? Well, in his 'Reflections on my critics', Kuhn does accept that there is a minimal sense in which he is willing hereabouts to be regarded as a 'Relativist':

> [There *are*] contexts in which I am wary about applying the label 'truth' . . . Members of a given scientific community will generally agree which consequences of a shared theory sustain the test of experiment and are therefore true, which are false as theory is currently applied, and which are as yet untested. Dealing with the comparison of theories designed to cover the same range of natural phenomena, I am more cautious. If they are historical theories . . . I can join with Sir Karl [Popper] in saying that each was believed in its time to be true but was later abandoned as false. In addition, I can say that the later theory was the better of the two as a tool for the practice of normal science . . . Being able to go that far, I do not myself feel that I am a relativist. Nevertheless, there is another step, or kind of step, which many philosophers of science wish to take and which I refuse. They wish, that is, to compare theories as representations of nature, as statements about 'what is really out there'. Granting that neither theory of a historical pair is true, they nonetheless seek a sense in which the latter is a better approximation to the truth. I believe nothing of that sort can be found.[18]

'Partial communication' is about speaking at cross-purposes,[19] speaking from bases which are not fully reconcilable with one another (and across which there is no point and no available *means* of making serious 'verisimilist' calculations). That is what – and 'all' – that Kuhn is really saying, here.

Linguistic idealism?

Sometimes the accusation against Kuhn (and likewise Winch, and Wittgenstein) surfaces in an apparently different form – not 'relativism', but 'idea-ism/idealism'. It is worth our treating it separately for a little while here – because considering the allegation of '(linguistic) idealism' allows for a perspicuous rebuttal of one of the main underlying worries that Kuhn's 'opponents' have: They fear that he is subject to a sort of mad credulousness. That he is committed to saying that things exist in the world if we *think* they do.

We are dealing with allegations which ascribe to both Kuhn and Winch doctrines they do not themselves embrace and they would and should disavow, but denying that they fall for an error does not ensure that they don't implicitly fall into it anyway. 'Linguistic idealism' is *an extrapolation* from their explicit arguments, but it is, unfortunately, a projection which involves placing a particular and *inappropriate* spin on their actual arguments. It is perhaps easy to see why the extrapolation seems plausible, even inevitable. The argument is that, according to their views (allegedly), the very concept of 'correspondence with reality' varies between cultures. Each is therefore as well justified as any other, each *in its own terms is valid – and there are no common or neutral terms in which they can be compared.* But doesn't this mean precisely, the argument against Kuhn and Winch continues, that the culture decides what is real, that it dictates what is true and false? How can it be otherwise if it is only in the terms of one or another system that we can say what is the case, and it is only in the terms of that system that we can decide whether what we say is the case is the case? Therefore, if that system says it is the case, then, logically, whatever it says *is the case.* Thus, if the witchcraft system says that someone is bewitched, *then (for them) they really are.* We may say that there are no witches, with (by the same logic) the consequence that, *in our system*, there are no witches. It does not follow from this that therefore there are *no witches at all* – that is, maybe there are, really are, witches 'for them' (but not 'for us'). Thus, it is the language of the culture, which dictates what can conclusively be said to be true or false, which decides what reality is like: if the culture says that witches are real, then the culture is correct, and therefore witches exist. For a Realist, of course, this is to get things upside down. It is how things are with the world that decides whether what we say is true or false, so it is whether or not there really are witches (and there aren't any) which dictates that when they say there are witches *they are wrong*.

Of course, in reality Kuhn *is not* arguing that the pre-Copernicans were correct about the immobility of the sun, nor is Winch arguing that there really are witches in Africa, though not in the UK, but this suggests why they are accused of linguistic idealism: they *seem to* be taking the view that reality is determined by our ideas expressed in language.

There is a major misunderstanding on the part of Kuhn's and Winch's critics. One of them, Bernard Williams, attempts to lay out the 'linguistic idealism' position as a clear series of logical steps.[20] Unfortunately he commits a major logical slip in doing so; but fortunately, because he attempts the logical presentation, this makes the error apparent and makes the absurdity of the alleged position transparent.

> (i)   'S' has the meaning we give it.
> (ii) A necessary condition of our giving 'S' a meaning is Q.
>   *Ergo*
> (iii) Unless Q, 'S' would not have a meaning.
> (iv) If 'S' did not have a meaning, 'S' would not be true.
>   *Ergo*
> (v)   Unless Q, 'S' would not be true.

Consider (iv) If 'S' did not have a meaning, 'S' would not be true. Consider the way in which 'true' is introduced into the argument. It is not as if the *conditions for the truth of* S were among the conditions Q. The formulation 'S would not be true' makes it sound as if S *is* true, and as though, therefore, *its truth* depends upon the initial clause of (iv): 'If "S" did not have a meaning . . .' This also gives the impression that (v) presents a challenge to the truth value of S. What is the alternative to S being true? That it is false. Thus, the conclusion which (v) seems to convey is that it is the set of prior conditions, Q, *which determine* whether S is or is not true.

Line (iv) is, in fact, an attempt to defy, rather than develop, the logic of the 'incommensurabilist' argument, which is quite plainly that the conditions Q are not conditions for *the truth* of S but conditions only *for its meaning*. The implication of the claim is that a necessary condition of S having a meaning is a condition of S saying something that is capable of bivalence, treated as a hallmark of an empirical proposition, one that is capable of being either true or false. The conditions Q are conditions for S being an empirical assertion, not of S's being a *correct* empirical assertion. One could accommodate this point by reformulating (iv) as:

If 'S' was not an intelligible empirical statement, then 'S' would not be the sort of thing that could be found to be either true or false.

But this is really, then, only a *repetition* of what is said in (ii) and (iii): it does not progress the argument, and, of course, it remains non-committal as to whether S is true or false.

But Q was only introduced as a condition for S to have a meaning. S's having a meaning is a precondition of S saying the kind of thing that could be true – that *would* be true if the evidence showed that what S asserted was indeed the case. If S is an empirical assertion, capable of being found either true or false, then of course *the meaning of S does not establish its truth*: empirical inquiry is required to determine whether it is indeed true. Whether Wes is more or less than six feet tall gets settled by measurement, not by the mere assertion that 'Wes is under six feet tall.' Q is a precondition for S meaning what it does, and, again, of course, whether S is or is not true does depends on its meaning what it does, but only because that tells us what kinds of facts would be relevant to determining its truth. However, this does not mean that in any usual sense of the expression, Q includes the condition for S being true. Q is not among the things that we can cite in support of S, but only a basis on which we and others can know what we are saying when we declare S.

Thus understood, (iv) and (v) are seen as repetitions of each other, variant expressions of the claim:

If 'S' did not have a meaning, we couldn't even say 'S', that is, couldn't say that it was true or false.

Once more on (iv), which now shows the depth of the confusion involved: S changes its identity in the course of this assertion. If S expresses the claim 'Wes is under six feet tall', then S does have a meaning. It means that Wes is under six feet tall. That S depends on condition(s) Q to have this meaning does not detract from the fact that S does mean: Wes is under six feet tall. Under other conditions than those that obtain, under conditions which comprise the absence of Q, then there is no S. Under condition(s) non-Q, either the words 'Wes is under six feet tall' are meaningless, or they mean something other than what they mean under condition(s) Q. In either case, a string of sounds or even a statement which does not mean that Wes is under six feet tall just is not S.

Much more concisely: the arguments of (i) to (iii) are for Q as an identity condition for S; (iv) and (v) either reiterate (i)–(iii), and are superfluous, or they distort the meaning of (i)–(iii).

Williams's exposition then makes it sound as if *among the conditions Q* are those which not only make it possible meaningfully to assert S but also those which decide whether S is true. We can see from 'Wes is under six feet tall' that it is Wes's height which is the state of affairs which determines whether this S is true, for this is involved in our understanding of the sentence 'Wes is under six feet tall.' The conditions Q certainly dictate that the thing to do is to measure Wes's height in feet to determine how tall he is, but we cannot determine from that fact about Q alone whether Wes is more or less than six feet tall! Nor can we even presume from the occurrence, abstracted from any context, of the sentence 'Wes is under six feet tall' anything about the world. The string 'Wes is under six feet tall' might be a hypothesis, something to be ascertained by inspecting the state of affairs referred to, calling for measurement. On the other hand, 'Wes is under six feet tall' might also be a description of the state of affairs – if it is a *report back* on the (proper, competent and accurate) measurement of Wes then it does describe the state of affairs, provides the evidence that confirms the hypothesis. The *meaning* of 'Wes is under six feet tall' depends on the way it is advanced, whether it is a hypothesis as to what measurement will show, or a citation of what the results of measurement were, or a guess, or (etc.). The importance of this difference will be obscured if we keep thinking of things as being either 'inside' or 'outside of' language (or of sentences (let alone parts of sentences), meaning independently of their context(s) of use). Making measurements and reporting their results is not really adequately described as a linguistic event or an occurrence 'within language'.

In short: the linguistic idealism which Williams presents is a straw position. It is not a position which could be put forward without basic and deep error. Kuhn and Winch make no such error – their view, as usual, is much more 'modest' than foes (and friends alike) suspect. They spent much of their careers after their early 'successes' trying to explain this to people. We hope that our readers at least are in a mood to listen, and read, and see – and not presume that Kuhn (or Winch) is committed to something 'exciting' and *absurd*.

A limit to understanding

Far from arguing that we have to accept that the convictions of the pre-Copernicans or those of believers in magic are true, Kuhn is in fact arguing that we *cannot possibly* make their way of thinking our own. There is *a* sense in which we sometimes use the word 'understand' to mean something like 'potentially be seriously able to say, myself'. Kuhn does not think it is an aspiration of the historian of science for what they are describing (that is, old science) to come meaningfully out of their own mouths – Kuhn is not preaching the Ptolemaic system!

The kind of understanding that Kuhn and Winch seek to provide us with is one which allows us better to appreciate the very-other's point of view, but which leaves us a *very* long way short of accepting their point of view, *further perhaps than before we read Kuhn and Winch.* This aspect of their philosophy, this respect in which they are the true 'anti-Relativists', is utterly missed by most commentators, pro or anti. Kuhn and Winch aspire to present the point of view of 'the Other' to us in such a way that we can understand how that point of view could be compelling to them, how, *in those same circumstances and under those conditions,* we ourselves would have thought (and been utterly and unquestioningly comfortable in doing so) in exactly those same ways – and how, in the circumstances in which we actually are, those other/older ideas *are complete non-starters.* Putting aside all that we now know from post-Copernican astronomy, and imagining ourselves gazing into the skies, reading the astronomical tables, immersed in Aristotelian physics, and so on, can we not see how the idea that the earth was a fixed and central point would have seemed wholly convincing, supported not least by the stark obviousness of what our own eyes could see. But, of course, this is an almost infinitely long way from suggesting that it *could* be convincing again, that we could seriously put aside our post-Copernican culture and unquestioningly embrace the Ptolemaic conception. The conditions Q that enabled the Ptolemaic system to have the sense that it did have, for its constituent assertions to be put forward as serious, even confirmed hypotheses, have been comprehensively eroded. We can recite the lines from the Ptolemaic system, but we cannot, ourselves, deliver those lines as stating a series of serious hypotheses that anyone should attempt to test, let alone as a series of assertions. Here *is* a limit to the extent to which expressions from 'alien' schemes can enter into, can be translated into, our 'own language'. They can only enter as things that we can say in, so to speak, quotation marks: we are only *repeating*, without endorsing, the things that they – the witchcraft believers, the pre-Copernican astronomers – say. There is no affirming these remarks as *expressions* of our own thought.[21]

In sum, everyday language users or (especially) scientists 'live' in one paradigm at a time (within any one given field): a contemporary physicist for example cannot seriously make *assertions* in 'Newtonese' – 'Newtonese' is, according to Kuhn, more dead than it is according to a 'point of view' relativist, *or* than it is according to a cumulativist about scientific knowledge (such as a Logical Empiricist).

So, for example, an Einsteinian scientist who makes a good (by Kuhnian standards) study of Newtonian physics, who has understood it fully *in Newtonian terms*, will then be able to say with fluency only 'What a Newtonian would say in this case . . .': our modern physicist would not be able to say these things on his or her own behalf, in the voice of 'What *I* would actually say here is . . .' This is what Kuhn calls 'irreversibility' in the development of science.[22] The point of the whole exercise is, of course, to be accurate about what could and should come out of the Newtonian's mouth, and the nature of the difference between that and what does come out of the Einsteinian's. Kuhn is completely *not* saying, 'From a certain point of view, the old paradigm is still true.' He expressly holds, as we have made clear, that Newton can be rendered a special case of Einstein (and thus kept true) only at the cost of *misunderstanding* Newton's theory. There is a real *clash* between paradigms for Kuhn, in a way typically ignored or finessed by relativist readings of him, at the cost of their completely missing Kuhn's point.

But the clash is *not* the clash of P and not-P, either. For us as Wittgensteinians, rather than followers of Quine or Davidson as such, it helps to point out the large extent to which what Kuhn is really talking about is ways in which changes in paradigms engage in and constitute the reconfiguration of what Wittgenstein sometimes termed 'grammar'. For us, language is not something that floats free, but something which is thoroughgoingly interwoven with activities. 'Conceptual change' is integral to change in ways of organizing activities: what words can mean depends on their connection to, and part in, our activities. When Wittgenstein says that to describe a language is to describe a way of life, he does not, as Sapir-Whorf would have it, mean that the ordinary meaningful use of some particular language is in the end incomprehensible to someone thinking or speaking in the 'different ontology' of another language. That way, clearly, lies self-refutation. What *does* Wittgenstein mean, then? Let's bring it right down to basics. Take the word 'king'. The word 'king', in the English language, is the name of a chess piece and also the title of a monarch – and it is clear that republicans can use and understand the word 'king' in the latter sense, in a perfectly ordinary everyday way, just as well as monarchists.

The disappearance of an activity does not deprive us completely of the use of the language it is interwoven with – Latin teachers would have been living some big lies, otherwise; and though the inhabitants of a republic cannot themselves swear allegiance to a monarch, it is not impossible for them to understand what it is to do this. Is sum, Kuhn, like Wittgenstein, is on about activities – linguistic activities *intertwined with* non-linguistic activities in *various* ways.

Some helpful analogies for incommensurability: philosophy, literature, religion

By this point, we hope to have satisfied and convinced some readers. But we suspect that many readers will still be feeling certain nagging doubts about what it all means. (Perhaps *especially* if they know the passages where Kuhn gives precision to notions of incommensurability in *RSS* etc., which can seem complex and daunting.) Kuhn is still not necessarily that easy to read, even after our 'help'. Are there further useful analogies at hand, to communicate the general character of what Kuhn is up to in *SSR* and *RSS* vis-à-vis 'incommensurability'?

Well, for example, *philosophers* read each other's books, but often they just don't *get* what the other is saying. The characteristic form of intersubjective philosophical puzzlement is perhaps not to know one's way about in someone else's discourse; for its components to seem perfectly familiar, but for the upshot to be uncanniness, or a wild oscillation between the feeling of incomprehension or of unbridgeable difference, the sudden dawning of mutual understanding, and then a rift opening up unexpectedly again. *Philosophical disputes are quite commonly (as have been those around Kuhn) more over what the parties are respectively trying to say, than about how things are.*

And here, in a way, we have of course come full circle right back to Kuhn's earliest efforts to describe what a scientific revolution is. Namely, an upheaval in the course of which the 'discipline' of normal science – its characteristics of having a regularity and a cumulative nature not possessed by (for example) philosophy or sociology – is partially suspended, and instead a situation is temporarily found in which philosophical-type disputes are possible. Moreover, such an upheaval specifically involves such-and-such becoming a candidate for truth-or-falsity when it didn't even make sense before, or vice versa. So perhaps what we are saying should be unsurprising after all. When Kuhn is talking about incommensurability, a really apposite analogy then may be *to the situation of two persons or schools locked in ongoing philosophical misunderstanding.* And again, such philosophical puzzlement, where what one side says must seem either pointless or nonsensical to the other, is virtually never resolvable through semantic clarification, through a simple mutual demonstration of meanings. (The philosopher must practise a 'hermeneutic' effort to understand another philosopher – and roughly this is what is called for in the philosophy of science, too.)

As we have already noted, Kuhn himself tends to use analogies to *literature*. Particularly useful, we think, is thinking of poetry (both the 'understanding' of poetry in general and the further issues arising in its translation), where words are not *transparent* – their whole point is often to fit in particular ways with the lines that surround them in the poem, or at least to facilitate specific kinds of associations and connections which are, as it were, provided for in the language. Take for instance the opening two lines of Robert Frost's 'On a bird singing in its sleep':

> A bird half wakened in the lunar noon
> Sang halfway through its little inborn tune.

Isolating one of these lines from the other would make a hopeless nonsense of their poetry. And even were they so isolated, then we could still say that the non-transparency of the language here is shown line-by-line, by for instance the importance of the rhyme of 'lun-' with 'noon', the aural-semantic 'allusion' from the word 'noon' to the concept 'moon', and the powerful metaphor of 'lunar noon' itself. None of this kind of specificity is present in the words, 'Pass the salt' (in almost any imaginable ordinary language context for them). In short, words in literature are typically *tied together* in ways analogous to the interconnections Kuhn sees as crucial within scientific lexicons. The incommensurability of lexicons, then, is like the untranslatability (and, indeed, similarly, the unparaphrasability) of literature.

Finally, compare Wittgenstein's comments on *religion*:

> Suppose that someone believed in the Last Judgement, and I don't, does this mean that I believe the opposite to him, just that there won't be such a thing? I would say: 'Not at all, or not always.' . . .
>
> If someone said, 'Wittgenstein, do you believe in this?' I'd say: 'No.' 'Do you contradict the man?' I'd say: 'No.' . . .
>
> Would you say: 'I believe the opposite', or 'There is no reason to suppose such a thing'? I'd say neither.
>
> Suppose someone were a believer and said: 'I believe in a Last Judgement,' and I said: 'Well, I'm not so sure. Possibly.' You would say that there is an enormous gulf between us. If he said 'There is German aeroplane overhead,' and I said 'Possibly. I'm not so sure,' you'd say we were fairly near.
>
> It isn't a question of my being anywhere near him, but on an entirely different plane, which you could express by saying: 'You mean something altogether different, Wittgenstein.'
>
> *The difference might not show up at all in any explanation of the meaning.*[23]

The interesting thing in each of these cases – philosophy, poetry, religion – is that understanding can be really difficult, and we might sometimes even find it *impossible*, even though the words involved are all ones in a language that we know. Thus it is that Wittgenstein's difficulties in understanding what religious people might mean by claims to believe in a Last Judgement applies to a claim in which each word is a perfectly ordinary one *when used outside the religious context*. 'Translations' are of course nevertheless made; but there is a sense in which we take them to be inevitably 'partially successful' (cf. 'partial communication') – there is something, which does not need to be strictly semantic, which is generally missing in even the best such translations. This is what Kuhn had in mind, we believe, when he spoke of what Quine and Davidson ignore – what gets lost in translation.

That there are real difficulties in understanding across different cultures and in translation between them does not come as a surprise to us with respect to philosophy, poetry and religion . . . so why should there be any difficulty in accepting that the same is true in the case of (history of) science? Unless, of course, one thinks that it is somehow demeaning to science to link it with philosophy, poetry and religion – but is that anything other than a prejudice? Just as in the humanities cases, then, perhaps, if one is to understand and depict a vanished scientific tradition, one will need to find novel and surprising and uninevitable ways of rendering it into terms that we can do *something* with. Understanding past science will be more like producing good paraphrases or translations of poetry than like producing good translations of car maintenance manuals. Here is Kuhn, in *The Road since Structure*:

> the *'difficulties in communication' arise between members of different scientific communities*, whether what separates them is the passage of time or the different training required for the practice of different specialties. *For both literature and science . . . the difficulties in translation arise from the same cause:* the frequent failure of different languages to preserve the structural relations among words, or in the case of science, among kind terms. The associations and overtones so basic to literary expression obviously depend upon these relations. But so . . . do the criteria for determining the reference of scientific terms, criteria vital to the precision of scientific generalizations. (238; emphasis added)

What is an ontology?

To sum up what we have been saying here about Kuhn's view: Words cannot be compared with reality. Nor can concepts. Nor even, in *any* direct respect, can sets of these – 'conceptual schemes', or, indeed, ontologies.[24] Basic ontologies ('furnitures of the universe') cannot be compared with reality, with nature. We would better say that only scientific claims can (*in a way*) be (for instance, *through* being tested 'intra-paradigmatically', and occasionally *through* being compared and contrasted with other dying or incipient paradigms).

'Isn't an ontology itself regardable as making a claim?' No – no one *says* ontologies (except, sometimes, metaphysical philosophers). Ontologies are what one works with; one's resource, not one's topic. Ontologies are, if you like, what make claims possible.

An ontology is at best specifiable as something like a list (such as 'matter', 'form' . . .) of properties that things (using that word now in the vaguest possible sense) can have or cannot have or have to various degrees, etc.

Scientific claims are precisely makeable only 'within' one paradigm or another, on the basis of one ontology or another. Advocates of different paradigms thus misunderstand each other, in roughly the respect in which philosophers find themselves not infrequently systematically puzzled at things that other philosophers say.

But what exactly is an ontology, now? If we repudiate on Kuhn's behalf the untenable claim that ontologies represent or 'correspond' to hermetically sealed systems of meaning, what do they 'represent'? How are they to be understood?

Outrageous as it may sound, we think that Kuhn's point of view, what he is trying to say, can be fully appreciated only if one gives up the notion that what it is for one to understand an ontology – a 'lexicon' (in Kuhn's terms), a particular scientific 'conceptual scheme' (in the phrase commonplace in philosophy) – is to do anything like grasp a metaphysical system, or comprehend a set of superfacts. Instead, one should think of understanding an ontology, roughly, as 'getting' a style. We are not saying simply that there is an analogy between ontology and style. No: we are saying that one should try to see an ontology *as* a style, as (a) form; and that that is the only way one *can* see ontologies in the present context, without falling into philosophical illusion. There is no such thing as magisterially surveying ontologies from an external – semantic – point of view. There is no such thing as an external point of view to language, or indeed to our best science, which alone could provide one with a 'position' from which to entertain the alleged claims of 'Metaphysical Realism' *or* 'Relativism'. Only if there *were* could it make any sense straightforwardly to give the true account of the history of scientific progress towards 'cutting nature at its joints' ('Realism') *or* to describe incommensurable systems of meaning ('Relativism'). The latter is self-refuting; the former misses (or accounts purely reductively and inadequately for) the very difficulty which prompted Kuhn's philosophy of science in the first place, the 'strangeness' of much past science, if one actually looks at it properly, rather than merely propagandistically.

The difference between chemistry before and after the chemical revolution, then, is *not* that the meanings of the sentences of phlogistic chemists are literally inaccessible to us. It's not that *their* ontology is literally 'cognitively closed' to *us*. It's rather that they operated with a significantly different taxonomy – a taxonomy which is worth calling 'significantly different' because it is directly translatable into ours only at the cost of a fatal (to historical understanding) loss of style, of form, of *sensibility*. Kuhn wants us (as historians and philosophers of science) to put ourselves more closely in tune with the *connotations* of the words of the old paradigm. If we 'translate' ignoring these, we're not really translating at all (in the 'full' sense of the word . . .). We're *just* producing a version of them in our own terms – *that* is the upshot of following Davidson (or Quine, Rorty, Kripke or Putnam. These are just different kinds of Whigs). Being able to translate, so as to understand merely from our own unalloyed and unnuanced and unaffected point of view what the sentences of the old scientists meant, is not enough.

What is needed is to be able to get much closer to understanding what is lost in such a 'translation', much as in the situation of the translation of natural language, or particularly of poetry. What is lost is not best described, we think, as a semantic item, a meaning. It is rather the awareness – the 'presence' – of the systematic network of associations and connections of words with one another and with certain practices in Newtonian physics, and *the 'grammatical effects' of that system*.[25] It is something like a feel, an atmosphere, a style (of reasoning, of writing). To think 'mass' or whatever *à la* Newton is to be able imaginatively to 'recreate' that atmosphere, and to regrasp those (linguistic) effects.[26]

Some may still find the above uncomfortably vague. An ontology – that seems a substantial, manly thing. A thought-style,[27] an atmosphere, a form of language – that may seem in comparison flaccid and hand-wavy. But we hope to have provided already the *material* with which to undercut these impressions. One is under a deep illusion if one thinks that fundamental ontologies can be spoken, can be simply declaratively given, at all. They can 'only' be understood as, roughly, *ways of speaking*. Though they are not like Rortian 'vocabularies', for Kuhnian 'taxonomies' are far more structured than that (and so, of course, our analogy to literature is only . . . an analogy).

These definite and 'structured' ways of speaking, these taxonomies, *must* be understood in a wholistic fashion. Kuhn's wholistic incommensurability, especially in the later formulations, applies to sets of interdefined terms. If you change the locus of one term in relation to other terms, you change (the use of) that first term also (when they are interdefined) – in a different theory it's no longer the same term. Hence, the translation can't be effected on the one

hand, and, on the other, in so far as the terms *are* different and conflicting, then of course we can't *affirm* equally statements featuring
each of them.

This point is terribly important; let us elaborate and exemplify it
a little. Words/terms in a scientific context have an extremely specific web of connections with the other words/terms around them.
Take a symbolic generalization – that functions as a tool, as a basis
for a number of crucial exemplars (exemplary problem-solutions) –
like Newton's 'F = M.A'. Kuhn (and, somewhat similarly, Feyerabend too) argued that this is only the equation that it is if it stands
in very specific relations to various other equations, and also to
various laboratory procedures, and so on. One might say that it is
only the equation that it is in a certain great physics text (Newton's
Principia), and in other texts which have stood in a particular historical relation to that text. It is in a way absolutely tied to its scientific context. By contrast, an ordinary language expression, like
'Pass the salt', is much less context specific. It can be used in numerous settings, and can survive many other changes in the language
around it without having to suffer any significant modification in
its use.

Now, in a way, lines of poetry are less context specific still. They
are endlessly recontextualizable. Thus they might *seem* the absolute
opposite of scientific terms.

But the analogy we have been pursuing (with literature, following Kuhn) still has force. For the web of connections a line of poetry
has with the language around it actually has a very great deal of
specificity, even if it is far less formalized than in the case of scientific terms. Unlike ordinary terms, words in poetry are not *transparent* – their whole point is often to fit in in particular ways with
the lines that surround them in the poem, or at least to facilitate specific kinds of associations and connections which are, as it were, provided for in the language. Unlike in the case of most ordinary
language, literature shows clearly its 'wholistic' nature – a wholism
shared, according to Kuhn, with scientific language and practice.

Conclusions

Let us close this chapter with one further linking remark to
Wittgenstein. We have endeavoured to rebut the almost invariably
wrong/overplayed attributions to Kuhn of 'relativism'; *and* to
present the sense in which Kuhn can intelligibly hold on to

something which one could if one wished choose to call, say, '(minimal conceptual) relativism'. The latter is in the conceivability (and actuality) of a sense of 'concepts' in which it is intelligible to speak of different concepts. The most crucial passage of all on this in Wittgenstein himself runs as follows:

> I am not saying: if such-and-such facts of nature were different people would have different concepts (in the sense of a hypothesis). But: if anyone believes that certain concepts are absolutely the correct ones, and that having different ones would mean not realizing something that we realize – then let him imagine certain very general facts of nature to be different from what we are used to, and the formation of concepts different from the usual ones will become intelligible to him.
>
> *Compare a concept with a style of painting.* For is even our style of painting arbitrary? Can we choose one at pleasure? (The Egyptian, for instance.)[28]

Our concepts are not, Wittgenstein reminds us, 'fixed' by nature. 'Our' style of painting – or of writing, for example of writing poetry, with the distinctive kinds of effects and 'aspects' that that yields – is not as a result *arbitrary* – and no more, of course, is our style of science, our scientific sensibility. One might talk here of the needs of our lives and the shapes of our concepts – one might talk, that is, of 'form of life'. Kuhn is with Wittgenstein in seeing nature not as a standard of and for comparison, except as 'within'[29] a paradigm. Only in 'intra-paradigmatic' thinking does nature play a precise and demarcatable role. Kuhn is concerned that we will do bad history and philosophy of science if we – absurdly – take our concepts to be 'absolutely the correct ones'.

In a nutshell, Kuhn is interested in *real* cases where the formation of concepts different from ours has happened, has been the case. Kuhn points to the same process as Wittgenstein, only in very concrete circumstances: in scientific revolutions. Kuhn sees scientific revolutionaries as '[imagining] certain very general facts of nature to be different from what we are used to', and thus as finding themselves needing '[to form] concepts different from the usual ones'.

In conclusion, then. Perhaps the most important thing we hope to have achieved in this chapter is not just to have explained what incommensurability is, but to have got over to the reader *the full 'triviality'*, moderateness, reasonableness and 'descriptiveness' of the concept. So long as one finds, as we find, the clear preponderance of Kuhn's writings leaning in the direction we have indicated, then one finds no reason to judge Kuhn guilty of extremism or self-refutation here. On the contrary, what Kuhn means by 'incommen-surability' ought to be recognized precisely as a to-be-expected aspect of what is still we think very much worth calling 'the growth of scientific knowledge'. The philosophers who have argued other-wise have, in the main, failed to understand the nature of Kuhn's contribution.

Though we nevertheless return to some of their concerns once more, when we take up that expression which most confused Kuhn's philosophical and scientific critics in the first place: 'world changes'.

5

Incommensurability 2: World Changes

There is no getting away from incommensurability, and in this section we deal in proper detail with one of its most famous consequences, the idea of 'world changes'.[1] Through that topic we deal with perhaps Kuhn's main explicit philosophical affiliations: first, his persistent Kantianism, and then his more recently accented Darwinism.[2]

On our first encounter with the idea of 'world changes' we argued that Kuhn's formulation of it, understood in relation to his historical accounts, and understood as using the expression 'different worlds' in a routinely idiomatic fashion, could be construed as a harmless restatement of the case that Kuhn had already made. At the same time, however, we registered that this was not all that Kuhn wanted to do with the idea. He desired, further, to express a strictly philosophical rendition of the situation of the scientist undergoing major conceptual change – an understandable desire, but an unavoidably problematical one. He evinced some unease about the formulation(s) he was using, advanced the point throughout in a tentative way – and gave the impression of thinking that there was rather more to the idea than he had yet managed to either identify or express. This is a moment in Kuhn's thought that has definitely, we think, been troubled by the confusion of idiomatic and philosophical uses of the same word, namely 'world'. As used in expressions such as 'the world of high fashion' or *My World and Welcome To It!* (by James Thurber) there is little danger of one feeling that a difficulty is in the offing: one understands clearly enough what these sayings mean. In philosophy, however, the word 'world' plays a very different role, and operates as (pretty much) a synonym for 'reality', even 'reality as a whole'. Therefore, some of the temptation that Kuhn may feel to adopt and persist in a locution (world changes) that he admits he finds strange may precisely be because he feels that he is, in saying them, making remarks of strong philosophical significance.

Saying that scientists divided by their paradigms live in different worlds, and that paradigm shifts involve changes from one world to another can easily give the impression (especially if associated with all kinds of intellectual trends following from the 'linguistic turn' in modern thought: he alleges that 'research in parts of philosophy, psychology, linguistics, even art history, all converge to suggest that the traditional paradigm is somehow askew' (*SSR*, 121)) that something philosophically bold is being asserted, that 'reality itself' is different before and after a scientific revolution, that 'reality itself' has been changed by the revolution. And Kuhn has reason, in his own terms, to think that he must be seriously putting forward a philosophically consequential position when he advances such statements. We will argue that the inclination to think that philosophical issues are at stake is encouraged by (at least) Kuhn's commitment to his initial historical objective of doing justice to past scientists and by his strong inclinations towards a 'two moments' doctrine of perception/ observation, or in other words, to resort to a Kantian notion of 'phenomenal world'.

We will argue, however, that Kuhn's real difficulties originate in confusions about the relations between tenses, between different kinds of descriptions, and with difficulties in articulating different – especially historical and scientific – discourses.

Kuhn was not by any means alone in thinking that something philosophically fateful was going on. His commentators and critics have been only too ready to follow him in this conclusion, and 'world changes' has been conceived as a major issue in – if not the central and most damaging implausibility of – Kuhn's thought. However, Kuhn's uncertainty about his own meaning was genuine, and he was much less quick than the commentators and critics to decide what he must mean. Kuhn remained painfully dissatisfied with his own grasp on the idea of 'world changes' and made repeated attempts to reconstruct it. We will, below, examine a couple of the ways – both early and later – in which Kuhn grappled, and in radically different fashions, with this matter.

Kuhn's Realism

To get to grips with the question of why 'world changes' should be a problem that Kuhn felt he needed to take up at all, let alone to struggle with, we need to mention, first, Kuhn's Realism. Kuhn's most savage critics must surely include those who deem themselves advocates of 'Realism', and their attacks often emphasize that Kuhn's major fault is that he affronts their Realism – as Kuhn accepts he does, though not, perhaps, in the way, or for the reasons, that they imagine. It is said that his arguments express Idealism, not Realism, and are generally irrealist in character. Kuhn feels entitled, nonetheless, to insist that he, too, is a Realist, but not of the kind that regular Realists will find themselves comfortable with.

'Realists' are not all philosophers of a single stripe, and 'Realism' can include a variety of different ideas, the different kinds being perhaps best identified on the basis of the positions they see themselves as opposing – Phenomenalism, Instrumentalism, etc. One kind of Realism is adduced in response, precisely, to Idealism. The issue between those two is whether there is an external reality, one which exists independently of the human mind, with a character which is intrinsic to reality itself and distinct from (and only possibly coincident with) the categories of human thought. Realists insist that there is such a reality, accusing Idealists of holding that reality exists only in thought. Going so far, that is, as to maintain that reality is composed of those very categories: there is no difference between our minds and the reality they ostensibly attempt to apprehend. Kuhn's few words (in *SSR*) about turning from the way paradigms 'constitute science' to the way they 'constitute reality' might seem to license criticism of him for falling for the latter view, but, taken in relation to almost everything else he says, this cannot be right (see chapter 4 above). It is actually in the sense that a Realist is insistent upon a mind-independent reality that Kuhn is a Realist.

Here is latter-day Kuhn, once again attempting to distance himself from those apt to think of themselves as his inheritors, those who say that 'reality is a social construction.'

> [T]he world is not invented nor constructed. The creatures to whom this responsibility is imputed, in fact, find the world already in place, its rudiments at their birth and its increasingly full actuality during their educational socialization, a socialization in which examples of the way the world is play an essential part. That world, furthermore, has been experimentally given, in part to the new inhabitants

directly, and in part indirectly, by inheritance, embodying the experience of their forebears. As such, it is entirely solid: not in the least respectful of an observer's wishes and desires; quite capable of providing decisive evidence against invented hypotheses which fail to match its behaviour. Creatures born into it must take it as they find it. They can, of course, interact with it, altering both it and themselves in the process, and the populated world thus altered is the one that will be found in place by the generation that follows . . . what people can effect or invent is not the world but changes in some aspects of it, the balance remaining as before. (*RSS*, 101–2)

This is not, however, a matter of Kuhn belatedly coming to accept a point of view that his work had, up to this time, either pointedly or unwittingly undermined, but shows, rather, Kuhn insisting on a point that his work has, *throughout*, presupposed. Indeed, Kuhn's Realism is a crucial element in his particular brand of 'Kantianism', a brand that tries (like Hegel and various other post-Kantians) to dispense with the noumena (but cf. *RSS*, 104).

We have earlier tried to dispel the false impression that 'nature' has no part to play in Kuhn's philosophical interpretation of his historical account of change in science, insisting that the formation and change of paradigms might almost be seen as a collaboration between scientists and nature. Nature exists quite independently of our thought, and seems, on Kuhn's view, to be necessarily elusive of any definitive set of comprehensive categories (which is plainly not the same as: entirely elusive of all our categories). On the contrary, nature is indeed captured in scientific schemes, but only partially so in each case, and by many different kinds of categories. And nature can, of course, only be captured through one or another set of categories.

As is clear in the quote just given, and in our discussion of the *SSR* formulation of 'world changes', Kuhn does not come remotely close to advising us that natural reality is constituted by and transformed with our scientific categories. What, then, does he hope to gain by adopting and taking seriously this locution.

What is Kuhn's world changes problem, really?

We mentioned Kuhn's commitment to the initial historical objective of doing justice to past scientists, and this is a fundamental feature of his thought, albeit one to which, at moments, he can perhaps be too strongly attached. As such a fundamental feature, it can provide him with what he can see as one of the good reasons for persisting in this line of thought. In fact, doing so gives shape to the problem that Kuhn faces with respect to 'world changes'. This is manifest in the following quote from SSR:

> During the seventeenth century when their research was guided by one or another effluvium theory, electricians repeatedly saw chaff particles rebound from or fall off the electrified bodies that had attracted them. At least that is what seventeenth century observers said they saw, and we have no more reason to doubt their reports of perception than our own. Placed before the same apparatus, a modern observer would see electrostatic repulsion (rather than mechanical or gravitational rebounding), but historically, with one universally ignored exception, electrostatic repulsion was not seen as such until Hauksbee's large scale apparatus had greatly magnified its effects. (117)

Kuhn's root idea is that there is normally no good reason to postulate a difference at a basic level in the human qualities of different generations of scientists. Construe those qualities on the (minimal/'empiricist') assumption that what scientists report about themselves and others is usually what they experienced. Then conclude: if scientists then as now are equally honest, careful, scrupulous and the rest in their reports, and if they are reporting their experience, then their reports indicate/report what indeed they did experience. In that sense, no scientific observer's report (aside from the unusual, viz. dishonesty, carelessness, etc.) is invalid.

But then, given the assumption that there is only one and the same natural world to be observed, one is left with this puzzle: how can the observers of one and the same world each observe different things in it? This puzzle is most memorably stated in Kuhn's small thought-experiment of putting an Aristotelian and a Galilean together to observe the motion of a stone on a string:

> Since remote antiquity most people have seen one or another heavy body swinging back and forth on a string or chain until it finally comes to rest. To the Aristotelians, who believed that a heavy body is moved by its own nature from a higher position to a state of natural rest at a lower one, the swinging body was simply falling with difficulty. Constrained by the chain, it could achieve rest at its low point only after a tortuous motion and a considerable time. Galileo, on the other hand, looking at the swinging body, saw a pendulum, a body that almost succeeded in repeating the same motion over and over

again *ad infinitum*. And having seen that much, Galileo observed other properties of the pendulum as well and constructed many of the most significant and original parts of his new dynamics around them ... Do we need to describe what separates Galileo from Aristotle. .. as a transformation of vision? Did these men really *see* different things when *looking at* the same sort of objects? Is there any legitimate sense in which we can say that they pursued their research in different worlds? (*SSR*, 118–20)

He does not go on to answer these in the unqualified affirmative!!!

One world, two worlds, many worlds, too many worlds

The immediate temptation is to ask: what do they respectively observe, what is the content of their visual field? It is here that the second fairly fundamental and persistent assumption of Kuhn's comes into play, that of the 'two moments of observation/perception', or, as it is alternatively formulated, a doctrine of reality as both 'object sided' and 'subject sided'.

Kuhn, in *SSR* and for a time afterwards, thought about these matters very much in terms of visual metaphors – the idea of the Gestalt switch is clearly of that kind, and his thoughts around these issues were cast in terms of the content of the observer's visual field: what appears in it? He later came to regret both the Gestalt analogy and the visual thrust, regarding this as seriously misguided. To be precise: he regretted this analogy as an account of scientists' experience. He came to think that, though the 'Gestaltist' account works well of the historian (for instance, Kuhn himself), there is less of an analogy between the historian and scientists (who typically work in groups, and who experience conceptual change typically much slower than historians do in their imaginative journeys back into the past) than he had earlier implied.

The correct way to think about these problems (as we will see) is, for Kuhn on scientific change, in linguistic terms. However, in our view, the 'linguistic turn' provides a means of attempting to get himself out of a problem that had arisen in part because of Kuhn's residual attachment to an empiricist legacy, the idea that the issue turns on the contents of the visual field.

The doctrine of 'two moments', as we call it, draws on what is obvious, and obviously true. Differences between the observation reports of scientists from different traditions may be put down to differences in the traditions from which they come. In Kuhn's terms, what scientists will report themselves as having observed will depend on 'prior experience', but this merely means (at least in a gross way) that it depends on prior training, on the scientific scheme they have been trained into. Thus, Kuhn's 'thought experiment' can be set out: we know what the Aristotelian and Galilean respectively will report themselves as seeing – constrained fall and pendulum motion – because we know what the difference between Aristotelian and Galilean physics is. It is, we are saying, true but utterly unsurprising that the Aristotelian will report the first kind of motion, and the Galilean the second. The difference between them, then, really derives from their respective science.

We are suggesting, then, that it is not a good idea to think of the disagreement between the two (kinds of scientist) in the way that Kuhn sometimes does, as a perceptual difference, and as one which must therefore reside within the visual field; but this is how Kuhn has tried to envisage the commonplace about scientific difference we have just reiterated. Setting it up this way, Kuhn is drawn into thinking that the visual field itself is a composite, that what is in the visual field is *in part* put there by 'prior experience'. However, there must also be something in the visual field which is not a product of prior experience, for prior experience must be applied to something to generate an observation. And it is, of course, to an input from nature itself that prior experience must be applied.

Nature is present in the observations of science. Hence, vindication of our point that Realism has always been a fundamental, and from his point of view indispensable, element of Kuhn's thought. Scientists *do* observe the (one and only) natural world, but they do so through the mental preparation that prior experience provides, or in the terms that their conceptual scheme supplies. Equally, this makes plain why Kuhn wants to be both a Realist and a Kantian. Our knowledge of the world (not just our scientific knowledge) must always involve two elements: a contribution from the world, and a contribution from our minds, with the latter (so to speak) endowing structure upon the former.

Taking this together with the insistence on the integrity of scientific observers, then, Kuhn is propelled towards the view that there are two worlds. There is the real world out there, the world of nature, towards which scientists direct their observations. But there is also the 'phenomenal world' within which a scientist dwells, a world that is the object of his or her routine observation. (Though using 'two worlds' here, we might better – but less conveniently – say that Kuhn's is a doctrine of one-and-a-half worlds since the real and phenomenal worlds are not completely distinct. The phenomenal world is *partially* made up of input from prior experience, and it is also made up of a contribution from nature itself.) The doctrine of two worlds promptly generates, of course, a doctrine of many worlds, for scientists differ in their prior experience, and therefore the phenomenal reality varies, depending (grossly) upon the paradigm to which the observer owes allegiance.

These assumptions set the form in which Kuhn actually confronts the 'world changes' problem. World changes are clearly to be understood as changes in phenomenal worlds, the displacement of one phenomenal world by another. When paradigms change, phenomenal worlds then change with them. Scientists divided by a scientific revolution occupy different phenomenal worlds. Here is the remaining problem: how can it be that one and the same natural world can accommodate different phenomenal worlds? It is his inability to fully satisfy himself on this question that motivates Kuhn's returns to and repeated rethinking of it.

The given and the interpreted

Kuhn, like Sellars and many others in recent times, wanted to get entirely away from the empiricist notion of 'the given'. One of the ways that Davidson's attack on Kuhn can be seen to be dangerously uncharitable is in failing even to acknowledge this 'moment' in Kuhn's thought. That said, it is undeniable that, although to get beyond 'the given' is what Kuhn intended to do, it is not really what he succeeded in doing. Indeed, it looks as if his felt need to continue to speak of the 'contributions' of nature and of the mind to one's experience and to one's words makes it impossible for him to get beyond 'the given'.

Slightly more concretely: Kuhn's 'two moments' position is an uncomfortable one since, taken seriously, it is unstable on the question of whether or not the natural world does or does not manifest itself within our visual field.

At one level Kuhn's Realism involves him wanting to say something like this: there are lights in the sky which are the objects of astronomical observation, and these are manifestations of natural phenomena 'out there' regardless of what kind of astronomical doctrine we hold. Sometimes, Kuhn speaks as if ostension can straightforwardly give us this, even across paradigms. But what we see when we look at those lights (for we do see those regardless of our scientific allegiances) will depend on our prior experience, that is, on which astronomical doctrine we accept. This looks as though nature itself does appear – in the form of the lights which assuredly originate off the earth, however they do originate – in our visual field . . . but is to say that the stars and planets do not appear to the naked eye, and their identity depends on what we understand the lights to be, that is, which astronomical conception we contrive or adopt.

However, if we hold that whatever appears in the perceptual field can only appear in interpreted form, then, of course, whatever is 'out there' cannot actually manifest itself, cannot appear *as such* in the perceptual field at all but only in one or another interpreted form . . . and so Kuhn has this problem: how deep down does the interpretation go? And how is one to answer this latter question? Since what can appear in the field is an inextricable fusion of nature and interpretation, it is not possible to single out anything in the field which is the contribution of nature itself, however tempting it might be to appeal, as above, to the 'lights in the sky' example.

Kuhn appreciated, if not these specific difficulties, that there were difficulties, and his first attempt (in 1969) was to resort to a distinction between data and stimulus.

The natural world appears in our perceptual field as 'interpreted', as what are, thus, the data of science. Hence, the lights in the sky are data, external but interpreted – data which are produced in us by nature itself, and are the results of the upshot of natural effects: the causal input of nature is the 'stimulus'. The process of interpretation cannot, however, take place within the visual field, since what is present in the visual field can only appear there in interpreted form. Therefore, in a desperate manoeuvre, Kuhn seeks to argue that the interpretation takes place outside the visual field, and prior to the moment of visual perception. The 'process of interpretation' is then conceived as a causal process. It is in the interaction between the causal inputs of the external world and the causally governed responses of our nervous systems that the perceptual world is constituted.[3] Kuhn must then be bold. The obvious objection is that if we are confronted with an entirely natural and causal transaction between an external world and the nervous system, then presumably same cause, same effect. This would undermine the 'plurality of worlds' which Kuhn is, as we have explained, strongly tempted towards.

If the perceptual field arises from the natural interaction of the input from the world and of the nervous system, then the only way in which the two-moment thesis can be sustained is to have prior experience affect the nervous system itself: the nervous system is to be thought of as programmed by experience, and therefore can be reprogrammed by further experience; which really means, in practice, only that it can be either educated differently (as, in a scientific revolution, those following the new paradigm are trained differently from those brought up on the old one) or re-educated (as would happen if a scientist 'crossed the floor' during a scientific revolution, and came to think instead with the new paradigm). Kuhn never did report the studies he was making on computer simulation of these processes.

Kuhn has (merely) moved the line between given and interpretation rather than abandoned that distinction altogether, that is, has relocated the interaction between 'the given' and 'the interpreted' by moving it outside the visual field and into the interaction between the nervous system and the causal input of nature.

A second attempt: a taxonomic turn

Kuhn did not persist long in the previous way of thinking about the problem.[4] Already, in its terms, he had begun to prepare the way for what he conceived as a superior solution, namely, the adoption of a taxonomic solution.

In developing the idea of the re-education of the nervous system (*ET*, 310), Kuhn had sought to illustrate by an example of education – taking the case of a small child being given parental instruction on how to discriminate between different kinds of water-fowl on the basis of their observable characteristics. Kuhn's assumption that the way in which (scientific) language relates to reality is through the classification of the similarities and differences that constitute phenomena in their 'uninterpreted' state now comes into play. Scientific language, for him, consists in classificatory schemes. Therefore, it is entirely natural to envisage parental education as involving instruction in the acquisition, through observation, of a classificatory scheme. Kuhn's later development of the 'linguistic turn' is a matter, then, of dispensing with the arguments about data, stimulus and nervous system programming which have led to this kind of example, and focusing directly on classificatory schemata.

Kuhn is, in elaborating this example, also showing another influence on his thought, that of Benjamin Whorf and Edward Sapir, whose notorious Whorf–Sapir hypothesis held that the structure of language (in their minds, its grammatical structure) determines the nature of perception of reality. Whorf–Sapir denies that there is any necessary correspondence between the structure of any language and the actual structure of the reality that is perceived, and Kuhn pursues roughly this thought with respect to classification. Waterfowl possess an array of characteristics, and there is no reason to suppose that the selection of criteria by which one may differentiate one fowl from another is the only selection that may be made: the makers of classificatory schemes may make different selections from among the array of similarities and differences which an assortment of objects may exhibit (thus it is the characteristics which are – logically at least – prior to the objects: the characteristics which are selected to make up the classificatory discriminata serve to identify the objects). Thus, it is not only possible, but inevitable, that different classificatory schemes are possible for one and the same reality, and that different categorial schemes will therefore yield different objects to observe.

But do the different categorial schemata merely attach onto fully pre-existing objects or clusters?

Natural kinds

The key Kuhn paper addressing the issue just outlined is 'Dubbing and redubbing: the vulnerability of rigid designation' (in *RSS*). Kuhn's intent and fighting spirit is clearly evident in the title here. He signals that, unlike the 'essentialists', he will concentrate not only on the 'dubbing' (naming, baptizing) of kinds in the world, but on their redubbing, their reclassification – on revolutions, and their widespread nature and effects, in short. He further signals that he thinks that 'rigid designation' – the metaphysical idea at the heart of Kripke/Putnam essentialism – is vulnerable to Kuhnian objections, vulnerable to falling foul of what Kuhn has said about serious conceptual change in science (313)

The underlying neo-incommensurabilist 'position' which Kuhn wishes to defend and build upon is stated clearly at the outset of his paper. It is to be noted, once again, that it wisely, modestly and crucially focuses on the role not of the metaphysician but of the historian:

To understand some body of past scientific belief, the historian must acquire a lexicon that here and there differs systematically from the one current in his or her own day. Only by using that older lexicon can historians accurately render certain of the statements that are basic to the science under scrutiny. Those statements are not accessible by means of a translation that uses the current lexicon, not even if it is expanded by the addition of selected terms from its predecessor. (298)

Here is a pretty careful 'linguistic' (re)formulation of what we have suggested is an absolutely central message of Kuhn's, throughout his career. Kuhn is suggesting that incommensurability be understood primarily as the ('partial') 'untranslatability' of taxonomies of natural kinds across revolutions. The (double) scare quotes are necessary because 'Incommensurability . . . equals untranslatability, but what incommensurability bars is not quite the activity of professional translators. Rather, it is a quasi-mechanical activity governed in full by a [Quinean translation] manual' (299).

The crucial sentences, for the purposes of trying to see what Kuhn is up to in discussing incommensurability under the heading of 'world changes', are perhaps the following:

To possess a lexicon, a structured vocabulary, is to have access to the varied set of worlds which that lexicon can be used to describe. Different lexicons – those of different cultures or different historical periods, for example – give access to different sets of possible worlds, largely but never entirely overlapping. (300)

It is what Kuhn means to mean by such grandiose sentences as these that we shall now try to uncover.

As already suggested, the philosophical worries of Kuhn's mainstream philosophical critics, facing Kuhn's effort to establish the import of these changes of lexicon – radical alterations in taxonomy, in mode of classification or 'ontology' – might be expressed as follows: There is only one world, and the Realist (unlike the Relativist or Idealist) can have coherent things to say about our deepening knowledge of that (one) world, because we (all of us, especially via scientists) are always in touch with the world. We are in touch with it through our naming of bits of it, and our growing knowledge about the nature of what we name. It may that at different times people had radically wrong ideas about the nature of the world – they may have thought, for example, that water was a primitive element – but, through being in contact with bits of the world (such as water), and through naming it, they always had some kind of 'basis' to their claims. We can connect with them and what they said, because there are direct (albeit long) chains of connection – linguistic and (more generally) 'causal' – between their use of these words and our use of them. They may have 'meant' something very different by their words, but the reference of their words was just the same as our reference for the same words. Reference – and in particular a causal theory thereof – will settle the problem which meaning poses. Kuhn is wrong, because he thinks that incommensurability of meaning is important and deep. It is, in fact, completely shallow and unimportant, once one understands that the real reference for natural kind terms remains continuous over time and through 'revolutions'.

Kuhn made original and interesting replies to this kind of essentialism, making a strong case that the causal theory of reference will not solve the problems/challenges which 'incommensurabilism' poses.

The Kripke–Putnam essentialist story

The Kripke–Putnam essentialist story[5] is usually run in two main versions. It speaks of the dependence of meaningful and consistent language-use on continuity of reference across *space*, and also across *time*.

Reference across space

This is the famous 'Twin-Earth' case – Putnam alleged that the meaning of 'water' would be different in a place ('Twin-Earth') if and where what superficially appeared to be water was actually a different chemical ('XYZ' for short, that is, a substance with a different 'essence'). The heart of Kuhn's response is this:

> The terms 'XYZ' and 'H_2O' are drawn from modern chemical theory, and that theory is incompatible with the existence of a substance with properties very nearly the same as water but described by an elaborate chemical formula. Such a substance would, among other things, be too heavy to evaporate at normal terrestrial temperatures. Its discovery would . . . demonstrate the presence of fundamental errors [anomalies] in the chemical theory that gives meanings to compound names like 'H_2O' and the unabbreviated form of 'XYZ'. Only within a differently structured lexicon [from that of modern, post-phlogistic

chemistry], one shaped to describe a very different sort of world, could one, without contradiction, describe the behaviour of 'XYZ' at all, and in that lexicon 'H_2O' *might no longer refer to water*. (*RSS*, 310; emphasis added)

Is Kuhn merely fiddling here? Is he being a spoilsport, bringing in mere facts to bear on a philosopher's charming speculations? No! He is raising a fundamental difficulty for the essentialist/referentialist view. He is saying: 'If we try taking Putnam's example seriously it turns out that Putnam doesn't really offer up a tenable thought-experiment after all. Unless you are willing to fundamentally revise chemical theory, that is, to alter it in a radical way, to turn the paradigm – in the full sense of the words, the taxonomy of chemical theory – upside down.'

Bird counters on Putnam's behalf[6] that Kuhn is muddying the waters, that his (Kuhn's) objection can be finessed by means of switching to a time, perhaps the mid-nineteenth century, when we could know that XYZ is not water (and that water is H_2O), without knowing that XYZ cannot have the surface properties of water (unless modern chemistry is more or less completely wrong). But this just seems incoherent: it seems that now there no longer is a thought-experiment. For how are we supposed to know that water is H_2O if we cannot rule out that some water is XYZ, if we cannot rule out XYZ as a starter in the game? And we cannot rule out the latter unless (for instance) we know that XYZ cannot have the surface properties of water (unless modern chemistry is more or less completely wrong, is in need of a fundamental taxonomic overhaul). Unless and until one has a good scientific reason for insisting that water's extension cannot include XYZ, one will not need to endorse Putnam's 'externalist' conclusion. Kuhn's point is: such good reason is 'only' in fact given by the taxonomic etc. structure of modern chemistry. The apparent discovery of XYZ would prompt not a reflection on the alleged truth of referentialist/externalist essentialism, but would rather herald a scientific revolution in chemistry.

In short, Putnam's essentialism begs the question against Kuhn by ignoring the knock-on consequences of change in taxonomy, of change in the world-as-understood-by-science.

Time

Putnam suggests that 'water' referred to the same substance (H_2O) in 1750 as today, and that that's an end of the matter. Kuhn, by contrast, is interested in the structure of (any given) chemistry, and writes as follows:

> 'H$_2$O' picks out samples not only of water but also of ice and steam. H$_2$O can exist in all three states of aggregation – solid, liquid and gaseous – and it is therefore not the same as water, at least not as picked out by the term 'water' in 1750.[7] The difference in items referred to is, furthermore, by no means marginal, like that due to impurities for example. Whole categories of substance are involved, and their involvement is by no means accidental. In 1750 the primary differences between the species recognized by chemists were still more or less those between what are now called the states of aggregation. Water, in particular, was an elementary body of which liquidity was an essential property. (*RSS*, 311; emphasis added)

Kuhn then remarks: 'This is not to suggest that modern science is incapable of picking out the stuff that people in 1750 (and most people still) label "water". That term refers to liquid H$_2$O. It should be described not simply as H$_2$O but as close-packed H$_2$O particles in rapid relative motion' (312). Can the essentialist not acknowledge this point as a point about meaning, but still hold on to the essential-referentialist conclusion? No, because

> The lexicon required to label attributes like being-H$_2$O or being-close-packed-particles-in-rapid-relative-motion is rich and systematic. No one can use any of the terms that it contains without being able to use a great many. And given that vocabulary, the problems of choosing essential properties arise again . . . Is deuterium hydrogen, for example, and is heavy water really water? (312)

Essentialism can answer these kinds of questions only by ignoring the 'systematicity' that Kuhn is drawing our attention to, and instead arbitrarily counting certain things as essential, others as accidental – a procedure surely inimical to a scientific approach, and making a mockery of the idea of natural kinds. Kuhn's view is that we must understand the progress of science as periodically involving the junking of (at least parts of) taxonomies of natural kinds, and their replacement by others. 'Rigid' designation no longer seems quite to deserve the name – for, though scientific schemas do indeed (of course, and *contra* relativistic and sociologistic fantasies) describe/explain the world across time and space, they are themselves often destined, even in many of their fundamentals, not to have the permanency of what they are used to depict.

Bird again tries to counter Kuhn, by suggesting that Kuhn's reasoning does not touch the 'essence' of Putnam's conclusion that 'in all possible worlds water consists (largely) of H_2O'.[8] The word 'largely' here is interesting. Presumably, Bird is allowing what Kuhn also allows: that water with some impurities is still H_2O. But what, for example, if those impurities are XYZ? What if those impurities turned out to be somewhat more numerous than expected; what if about 25 per cent of water on the earth's surface were 'discovered' to be XYZ (so that water is still 'largely' H_2O)? Would essentialism still look so attractive, then?

It is only given the post-Lavoisier framework that we are forced to see water as largely H_2O. Absent that framework, 'water'-in-all-its-states is not necessarily viewed as a natural kind (as the quotes above from Kuhn make clear: liquidity was regarded as an essential property of water) – and still less is H_2O. Kuhn is bringing talk of possible worlds, one might say, back from its metaphysical to a more everyday (that is, everyday scientific) use. A taxonomy supplies a 'set' of possible worlds between which normal science goes on to choose. If something really threatens the taxonomy, we (imagining ourselves now into the position of scientists actually confronted by such an anomaly) cannot retreat to philosophers' assurances about what all possible worlds must turn out to be like. Rather, sometimes, we must face the need to uproot fundamental assumptions about the set of possible worlds available to us and enabled for us by our taxonomy, our 'ontology', our thought-style.

Redubbing is then at least as important as dubbing; and, of course, in concert with Kuhn's reasoned scepticism as to correspondencism, progress through revolutions is not well described as bringing us taxonomies which themselves come closer and closer to matching the universe's 'own' taxonomy.

Kuhn's conclusions

Kuhn's conclusions are as follows: '[D]ubbing and the procedures that accompany it ordinarily do more than place the dubbed object together with other members of its kind. They also locate it with respect to other kinds, placing it not simply within a taxonomic category *but within a taxonomic system. Only while that system endures* do the names of the kinds it categorizes designate rigidly' (314–15; emphasis added)

Part of the fundamental problem with essentialism then is to do with what we discussed earlier vis-à-vis the issue of incommensurability around:

- the anti-contextualist focusing on single words (such as 'water');
- the attempt to understand the reference of single words without any placement of them in an assertion and indeed in a broader context within which the assertion is intelligibly available; and
- the absurdity of 'point by point' renditions of 'translation' – recall the issue around translation discussed in chapter 4.

And part of the problem, relatedly, is the need for a still broader wholism, which we also touched on earlier: this is to do with Kuhn's sense that a taxonomy of terms must be taken relatively wholistically, and that that fundamental taxonomy/ontology inflects or is a whole style of thinking, which sometimes needs radical alteration, rather than supposed piecemeal alteration (of the kind which essentialists always present scientific change as constituting): 'Here and there the old and new lexicons embodied differently structured nonhomologous taxonomies, and statements involving terms from the region where the two differed were not translatable between them' (*RSS*, 315). We are really, here, in a new terminology, expressing Kuhn's constant insistence that 'conceptual schemes' are, because of their wholistic character, incommensurable, that one never really transplants an idea or expression from its context to a new one, for to remove an idea from its context is to denature it.

Kuhn's taxonomic conception has enabled him to defend and respecify the notion of incommensurability against the idea that it is reference, not meaning, that is important. While this respecification was needed and effective, it was not enough to satisfy Kuhn himself, who thought that there remained a need for a more substantive understanding of incommensurability, why it was necessary in the first place. A turn to evolutionary theory seemed helpful.

A third attempt: an evolutionary turn

An evolutionary parallel had been drawn in *SSR*, where the idea of the evolutionary tree as resembling, more, a proliferating bush had been called on to help dissipate the idea that for science to be evolving, it had to be evolving towards something. Much later, especially from 1990 on, Kuhn argued that the parallel needed to be made much stronger, that there was a great deal to be gained from developing his own ideas in the form of a Darwinian theory of scientific development. There are various ways in which this analogy can be elaborated. One main one was in construing Darwinism as a theory of animal speciation, and in recognizing that what Kuhn had been trying to say about the growth of the sciences was that that was akin to the proliferation of scientific species. The introduction of these new connections to Darwinism into Kuhn's scheme would lead to a retrospective change of emphasis: the main importance of paradigm shifts would be seen to be in the generation of new sciences, rather than in just the replacement of paradigms within the same specialism. It would provide an enhanced conception of incommensurability too. That could now be thought of as an evolutionary mechanism, one which, by ensuring discontinuity between specialisms, would provide an isolating mechanism, serving to separate each from each, and thus to give them their own, independent, distinct identities.

Around 1990, then, Kuhn was ambitious for his Darwinian project, which does seem to have a strongly naturalistic intent, not to be meant as just a metaphoric conceit; but a first use for his new ideas on this was to try to sort out the world changes problem, to make use of the idea of the 'evolutionary niche' as a restatement of, if not entirely a replacement of, phenomenal worlds.

Back, then, to the base problem:

Premise one: there is an 'external world' of nature.
Premise two: there is a plurality of 'phenomenal worlds' that scientists inhabit.
Problem: how can incommensurably discontinuous phenomenal worlds be accommodated within the one external world?
Answer: if we consider phenomenal worlds as 'evolutionary niches' then a 'diversity of phenomenal worlds' is simply subsumed into the now routinely accepted idea of a 'plurality of evolutionary niches'.[9]

Now to unpack this argument. One of the unattractive aspects of the notion of 'phenomenal worlds' for Kuhn must surely be that it invites subsumption under the standard dichotomy of 'objective' and 'subjective'. There *is* an 'objective world' of nature out there,

and there is, on the human side of the divide, a subjective – phenomenal – world. But to allow the matter to rest there is to expose
Kuhn to those critics who, for example, want to call him an 'Idealist' since to call the world according to science subjective is seemingly to say that it does exist *only* in the mind. Kuhn's resort to
evolutionary theory tries to preserve the thought, but to bypass the
dichotomy between the object and subject sides – in the same way
that J. J. Gibson tries to bridge the same dichotomy with the idea of
affordances. This was our point about one-and-a-half worlds, since
a phenomenal world is not a purely subjective one, being partially
constituted by the objective world itself. Therefore, Kuhn sees in the
aspect of the Darwinian analogy of evolutionary niches a way of
eroding the sharpness of the usual distinction between the subjective and the objective, the dichotomization of the phenomenal and
natural worlds. It will allow him to say that phenomenal worlds
are, themselves, real worlds:

> Can a world that alters with time and from one community to the
> next correspond to what is generally referred to as 'the real world'?
> I do not see how its right to that title can be denied. It provides the
> environment, the stage, for all individual and social life. On such life
> it places rigid constraints; continued existence depends on adapta-
> tion to them; and in the modern world scientific activity has become
> a primary tool for adaptation. What more can reasonably be asked
> of a real world? (*RSS*, 102)

Our own difficulties with this go back to our discussion of the
initial introduction of the 'world changes' idea itself. It is quite clear
that, from a historian's point of view, the contemporary scientific world which that scientist inhabits is, for that scientist, the
real world with which the scientists must contend. In that world,
the world of nature is treated as consisting in (say) two distinct
realms, that within the space between the earth and the moon, and
that beyond the lunar sphere. Within that scientific world the world
of nature is very different from the way it is in our contemporary
scientific world, for in the pre-Copernican world the nature of
motion differs between those two realms. Two associated emphases
of Kuhn's are important: on the collective character of such 'realities'; and on the categorical mode of scientific speech. The working
scientist's reality is the working world of the scientific community,
with its entrenched apparatus for conceiving, representing and
intervening in the world of nature that unquestionably is no matter
of individual subjectivity.[10] 'Phenomenal world' does, though, seem

to us a poor and misleading description for the characterization of the portrayal of the world of nature that is installed in any of these working worlds. Within that working world, its entrenched portrayal of nature has the status of reality, which is reflected, first, in the categorical mode of speech: that earth is not a planet is not a claim advanced in a provisional and cautious way. Harold Garfinkel (speaking for sociological purposes)[11] talks about facts as 'grounds for further inference and action', and means this in a definitional sense. From an observer's point of view, a 'fact' is that which, in the community under observation, is treated by its members as grounds for further inference and action (rather than for questioning, argument, etc.). In the same sense, the entrenched portrayal of nature in a scientific community is a reality, that is, it is that which, the historian determines, provided the grounds of further inference and action for scientific inquirers. Kuhn's resort to Darwinism thus seems to us a rather roundabout, and still rather ineffective way of trying to get these points across. Let us, however, look and see how Kuhn deploys the idea of an evolutionary niche.

An evolutionary niche is not a part of 'the world of nature' in the way that a tree or an atom might be conceived to be, but why should one deny reality to an evolutionary niche, say that because it is not part of the world in the way that a mountain is, that it is not part of the natural world at all? The existence of an evolutionary niche is, of course, creature dependent. A feature of the world such as that of a berry's being edible by pigeons is not one which pre-exists the existence of those pigeons. The feature only exists in relation to the species, and the evolutionary niche is thus identified in an interactional fashion – the evolutionary niche only exists between the world of nature and the species of creature, but, of course, given that relationship it is perfectly all right to say that the evolutionary niche exists in the world of nature. An evolutionary niche is no more a 'subjective' aspect of nature than is a mountain (after all, the mountain may figure in, be part of, the evolutionary niche) and is no less a part of nature than the mountain is, but it comprises a part of nature in a different way than the mountain does.

It is important to Kuhn, then, that the evolutionary notion of 'adaptation' be understood in terms of the notion of 'niche' rather than the other way around. It is, of course, common for Realists to insist on 'death and furniture' arguments. The idea that the world can be given meaning might not be wholly without merit, they may concede, but there is a limit to what can be done. The external world of nature is fixed, and sooner or later will inhibit whatever interpretations we may care to project upon it. We cannot walk through the furniture, and, more seriously, if we do not conceive of the world of nature in the right way, there will be fatal consequences – if we define something with the wrong chemistry as edible, then we will poison ourselves. We are forced to adapt to the external world. But Kuhn is questioning this notion of adaptation, treating it as too rigid and one-way, and therefore as providing a false dichotomy:

> Is it the creatures who adapt to the world or does the world adapt to the creatures? Doesn't this whole way of talking [about evolutionary niches] imply a mutual plasticity incompatible with the rigidity of the constraints that make the world real and make it appropriate to describe the creatures as adapted to it? (*RSS*, 102)

Kuhn's position is that what actually evolves is not, then, the creature in adaptation to a given and unchanging nature, for niches and creatures 'evolve together'. Returning to the scientific case, Kuhn is now licensed to claim that our 'worlds' are niches: 'Conceptually, the world is our representation of our niche' (*RSS*, 102–3).

One cannot make, in these terms, any trans-historical comparisons of phenomenal worlds, asking whether any one of them is better 'adapted' to natural reality than any other, for the notion of the niche is a two-way relation, and therefore one which is tied to the specific combination of environment and organism. One can't make comparisons of fit with other environments and organisms. The analogy, applied to paradigms, means that we can say that each is a reality in its own right, a relationship between the scientific community and its environment, and that we cannot say that one paradigm is better adapted to the same environment than others, since 'the environment' is now functionally defined, is a relative conception itself. Thus Kuhn dissociates the role that 'subject sided' and 'object sided' elements play in his own thought from that of the sharply separated contrast that they more usually make. If this movement were rigorously carried through, it would eventuate in the dissolution of a 'two moments' doctrine into a therapeutic reorientation of us back to our everyday (that is, everyday scientific) lives and concepts.

Kuhn's attempt to develop the evolutionary idea of the functional niche certainly looks like a kind of relativism of which he is often accused, insisting that each niche is as valid as any other, that the 'world' inhabited under any scientific paradigm is one which is just as real as any other. Kuhn's emphasis on the reality of these *phe-*

nomenal worlds is, in our view, a rather misshapen attempt to give further expression to his most basic and entirely plausible precept, that the difference between earlier and later scientists is not a characterological one. Earlier scientists are no less intelligient, industrious, scrupulous, visually well equipped, etc., etc., than more recent ones, and the differences between the earlier and later sciences cannot be put down to individual deficiencies on the part of those scientists that would affect their practical competence in science. In other words (and it is just this same point that Wittgenstein and Winch have agued with respect to 'primitive magic'), previous science is not rejected because its practitioners made some stupid mistake which is readily controverted by obvious facts that their successors have happened to notice. This last is no relativist doctrine, but neither is it most perspicuously expressed in the analogy with ecological niches.

In sum, the niche is not something which pre-exists the species that is adapted to it, but, rather, 'what actually evolves . . . are creatures and niches together' (*RSS*, 102). And, in favour of Kuhn's intent here, it should be said that 'niche' or (similarly) 'ecosystem' is an improvement over what Kuhn inherited; because it sounds less like an 'object' than 'environment' or 'world' does. It thus has a better chance of avoiding doctrines of 'the given' and of Scientific Realism *without falling into Idealism.*

Thus the real point of the evolutionary analogy is to bridge the opposition between the 'object-sided' and 'subject-sided' worlds. But perhaps a better way of making that bridge – or of showing, therapeutically, that the very idea of needing a bridge here is still too concessive – is available.

Our Rylean critique

We are unclear from the published work how seriously Kuhn intended his Darwinian turn. It could be an enriching of a helpful metaphor that he has already used, but it does look (in *RSS*) as though it is intended much more literally than that, and that Kuhn is about to set out on an ambitious and thoroughly naturalistic account of science as something that does actually (within the terminology of post-Darwinian biology) evolve. If this was Kuhn's intention then we find it entirely unpersuasive, and unnecessary.

Kuhn is attempting to break away from the empiricist philosophy of science that had been his inheritance, but such philosophical inheritances are deep, subtle and penetrative. Kuhn's need for Kantian thoughts, and therefore for phenomenal worlds, arises, we think, from a residual element of the empiricist tradition, namely the idea of the visual field as a location in its own right, and one which must possess contents. 'The contents of one's visual field' is a perfectly acceptable commonplace expression, but it is referring only to whatever one can see from one's current vantage point. Whether the traffic light is among the contents of one's visual field asks simply: is one now looking towards the traffic light? The capacity to identify things in one's visual field obviously depends on prior experience, as is clearly the case with knowing what a traffic light is, but one's capacity to make such identifications is not itself any part of the visual field. What is observed is not, therefore, any kind of admixture of natural and mental elements, and there is, therefore, no need to be troubled in the way that Kuhn is by the need to understand how such elements are fused in experience.

There is in fact available a ready way of disposing of Kuhn's puzzlement before it even starts, and thus of negating the felt need for bolstering one's view by reference to alleged naturalistic support for it (viz. non-metaphorical evolutionism): by means of Gilbert Ryle's distinction between 'thick' and 'thin' description. The difference between someone who sees 'a rain cloud' and someone who sees merely 'a cloud' in the same place is not a difference in the cloud that they see. Both can be said to see one and the same cloud, just as they can both be said to see something different – in this case, one can see something that the other cannot. It is not, either, that one can see one kind of cloud and the other sees a different kind of cloud. The one who sees the 'rain cloud' can tell, as the other cannot, what kind of cloud they are both looking at. The description 'cloud' is a thin description when compared with 'rain cloud' which is much thicker, that is, is replete with understandings that do not inform the thinner description. The one who sees the 'rain cloud' does not witness a cloud with additional features compared with the one who sees only the plain cloud, for the features of the cloud that the former notices are also within the visual field of the latter. But the one who sees the rain cloud can make more of what is seen than can the other, can notice things about this cloud that the other cannot. The difference in what they can respectively see lies not in the 'contents of their visual field', but in what they can bring to bear on their observation. The one who sees the rain cloud can discern the significance of features that the one who sees the plain cloud may not notice or understand. To say that it is a rain cloud is not to provide a mere report of an observation, but to involve a certain predictive element: it is a 'rain cloud' because it will bring rain soon, because it is being blown in this direction, will pass over high land, etc. The capacity to make such a prediction involves a deal of background understanding about how the size, colour, location of the cloud and the direction of the wind affect the behaviour of clouds in general.

Thus, there really need be no compulsion to find, as Kuhn does, something of a puzzle as to how the Galilean and Aristotelian sciences can both be looking at the same thing and 'seeing something different'. They are, of course, seeing the same thing *thinly described*, namely, the stone swinging on the end of a piece of string, but they are observing the swinging stone with respect to its motion. The motion of the stone is not to be given, for scientific purposes, by a thin description such as 'swinging back and forth', for they – the two scientists – differ over the kind of motion 'swinging back and forth' is, whether it is constrained fall or pendulum motion. These are two thick descriptions of the motion of the string, and presuppose the different apparatus of ideas that make up Aristotelian and Galilean physics respectively. The thick description of the stone's motion is not meant to provide only an accurate capture of the thinly described contents of the visual field for, of course, constrained fall and pendulum motion would each allow them to report in thin terms exactly the same back and forth movement. It is misleading to think that the physicists are engaged in attempting faithfully to describe their experience, to report accurately what they have found in their visual field, and to recognize that they are treating their observations as a means of applying specific scientific systems.

Their case is different from our 'cloud' example, where we wanted a contrast between the thinner description 'cloud' and the thicker description of the same thing ('rain cloud'). In the case of the swinging stone example, however, we are dealing with two thick descriptions. It needs, after all, to be recognized that 'pendulum motion' is in an important sense not, as it may seem to those of us who have inherited Galileo's system, a thin description at all, but one which depends on assumptions about what propels the motion, what would happen if certain counterfactual conditions obtained, and so on.

Being thick descriptions, because they are both paradigm based, Galileo's and Aristotle's are rival descriptions, and they cannot be treated as jointly compatible with the same observations. To intimate the latter would to treat the observation as though it were itself adequately described in 'thin' terms, but this is not so. There is no scientifically useful description which is neutral between the two rival descriptions. A neutral description would have to be at best something like: a stone swinging on a string. But to speak of 'a stone swinging on a string' is not to give a scientific description at all: it is naive, ordinary, and too basic to do anything with. Such a description would never or rarely be included in the report of an experiment. (The report, if by a Galilean, would run, 'A pendulum was set in motion; it . . .'). In other words, part of the contest between the two traditions is precisely over what it is that is being observed (which point Kuhn himself well appreciates in other contexts). The divergence between the two thick descriptions does not reside in the visual or perceptual field, for it does not reside in nature itself. It is the working world of science (not nature) that accommodates these different phenomenal worlds. Kuhn's difficulties with 'many worlds' can, we think, be put down to his confusions about the relations

- between different kinds of descriptions;
- between tenses;
- in the articulation of different discourses.

We have drawn out the way we see the first of these confusions, cashing it out in terms of 'thick' and 'thin'. What Kuhn himself teaches us is that the language of scientists-under-a-paradigm is, of its nature, categorical, and he also teaches us that paradigm shifts involve the retrospective eradication of a prior mode of discourse. What was categorically said under a prior paradigm can no longer seriously be asserted at all.

Kuhn's inclination to find an ongoing problem here, in so far as it is an inclination leading him to want a theory of meaning or a substantive metaphysics, is a result of the disregard of what he is otherwise teaching us in his own descriptive account of the course of scientific revolutions, which is that ways of describing things are withdrawn from us and retrospectively overruled (see for instance SSR, 114–15). We have taken it to be Kuhn's singular merit that he has sought to point out that the supposed problems of the philosophy of science are in most cases ones that the scientists must solve in practice. We might count this lesson as echoing some of Marx's arguments, and as paralleling those of Wittgenstein's philosophy or of ethnomethodological sociology.

Kuhn does not, however, appear always to appreciate that this is his crucial contribution, and occasionally falls into imagining that what the scientists are doing poses a philosophical problem in the classic sense: one that requires a philosophical doctrine for its solution. Thus, it is perfectly correct for Kuhn (the historian) to say that up until a certain date chemists identified only 'mixtures', and that prior to that date 'compounds' were absent from chemistry: therefore, 'first mixtures, and then compounds'. But this is of course a description of the change in chemistry's concepts, and has nothing to do with change in the nature of chemical substances. It is a very different thing to describe a historical change in chemistry's concepts than it is to engage in reconstituting chemistry's concepts, which is what those whose activities Kuhn is reporting on were actually doing. And it is another thing altogether to imagine chemical substances themselves actually changing their nature some time in the eighteenth century, coincident with the change in chemistry's vocabulary. We are not saying that Kuhn requires one to imagine the latter: but that the misinterpretations of him along such lines are at least understandable, given how long he persisted in making it appear as though there were a deep problem here which might yield to a theoretical solution.

The introduction of the concept of 'compound' was not intended to initiate the existence of some entirely new sort of phenomenon, one that exists as well as the phenomenon of mixtures. The concept of 'compound' was a rival to that of mixture (this is its interest to Kuhn, after all) and it is introduced so as to replace the latter (although, of course, the word 'mixture' remains in use in chemistry, but now, with a changed meaning, as a contrast class to 'compound'). If the revolutionary aspiration of 'compound' is successful, then the word 'mixture' (in its original use) will be withdrawn from further discourse in chemistry, except (in its new use) to refer to something quite specific and less chemically interesting (for example, soil being carried along in a river is a (physical) 'mixture' in this sense) – and one can date (though not to the very day, even in principle) the occurrence of that transition. Before this date, 'mixture'; for this period of years, the rivalry of 'mixture' and compound'; after this date, 'compound' only (plus merely physical mixtures). However, and this too is a descriptive observation, it is also equally clear that the introduction of 'compound' operates with retrospective force: it is not introduced to apply only 'from this date forward', but is to apply to cases that were adduced in the past as cases of 'mixtures'. In the new scientific discourse, these substances never were mixtures, they were compounds all along.

Of course, it is the historian's role to note and report such changes, not to endorse them, and the fact that for scientific purposes one discourse retrospectively pre-empts another is not something for which the philosopher need contrive a workaround.

There is only the appearance of a worrying divergence here between scientist and historian. What is necessary in the historian's terms to avoid Whig excesses is very different from what is involved in taking part in the installation of a paradigm, and Kuhn's phraseology of world changes here is really only a (striking) conceit. Says Wittgenstein, say what you like, so long as you don't blind yourself to the facts. As it proved, however, Kuhn's conceit proved a risky one, and served to confuse himself sometimes, blinding him to facts that otherwise are plain and pretty much indisputable. In sum, the 'world changes' locutions by no means sink Kuhn's project; but they are, we think, very needful (at least now) of being deflated down to earth. With our Rylean and Wittgensteinian manoeuvres, combined with the adducing of the morals of Kuhn's impressive discussions of kinds and of taxonomies (and of his somewhat more contorted discussions of Kant and Darwin), we hope to have put the reader in a position to understand why Kuhn wanted to say some of the things he said that have most bemused people – and why, in the end, he didn't really need to say them. Or at least: why one doesn't need to say them any more.

Conclusion

The Unresolved Tension

A cure for the philosophy of science?

Clearly, we believe that the real Thomas Kuhn has been less influential than the legendary one, something obviously frustrating to the real one, who rightly thought that the influences of legendary figures were often for the worse. Many of the supposed errors for which Kuhn has been condemned are not to be found in his work.

We have portrayed Kuhn as continuously concerned largely with one issue – spelling out the meaning of properly historical studies of episodes in the history of science for the philosophy of science. Late Kuhn came to think that his constant concern was best expressed as being with change of belief in science (*RSS*, 95–6, 112). Late Kuhn also came to fear that the emphasis on history had misled many, from whom he now needed to dissociate himself, and also that this emphasis had been excessive. He thought that as much could have been achieved with far less reliance on the historical work, accomplished on philosophical grounds alone. Kuhn was surely right that the emphasis on history has misled many. But does this mean that everything important in Kuhn's work could be deduced 'from [philosophical] first principles' (*RSS*, 112)? This was surely an overreaction. Kuhn demanded a lot from his readers, and the fact that they fell short of his expectations deformed, we believe, the shape of his later work. So concerned was Kuhn to avoid identification with 'historicizing' and 'sociologizing' circles – because of how he was misread in these circles, and because association with them made it difficult to be taken seriously by philosophers like

Shapere, Putnam, Popper and Quine – that he came to say that it was only philosophy – and indeed 'science' (biology, psychology, 'artificial intelligence') – that was his overriding interest. This seems to have been more the expression of a relatively recent view than of something that had always been true but that Kuhn had only latterly realized about himself.

A Kuhn loyalist might ask whether we ourselves have not misrepresented Kuhn by placing as much emphasis as we have on Kuhn's historical studies, mistaking the fact of their prominence in Kuhn's writing up to the mid-1970s for a sign that they were more important to Kuhn than they really were. We have urged the importance of Kuhn's historical studies as clarifying his philosophical arguments. We do not deny that anyone *could* get a clear conception of Kuhn's philosophical arguments from reading SSR alone, *if they read it carefully*, but it is equally clear from the last forty years that reading SSR alone commonly does give a woefully false impression of what Kuhn means. The historical studies show more clearly what Kuhn has in mind in his philosophical arguments, and help to moderate the anxieties that those provoked – the edifice of science is hardly shaken by the 'revelation' that Copernicus had much less responsibility for the revolution named after him than such naming might suggest, and that the scientific-technical merits of his work were less than he thought, and less advanced over previous work than he supposed. If *these* are the sorts of things that Kuhn is trying to tell us, then the exaggeration in the reactions to his work appears more clearly. Only to a historian (that is, not a practising scientist), and only then before they have done the required historical work, does a scientific revolution need to appear both real and *dramatic*.[1]

There is constancy of purpose throughout Kuhn's career, though the way it was pursued varied, especially after the publication of SSR. Thereafter Kuhn remained preoccupied with issues arising from the book, but did consider, and in his view significantly revise ideas expressed there, in his repeated attempts to elaborate, clarify, restate or defend his thoughts. Was Kuhn's thought ever quite consistent? There are conflicting philosophical strands in it, the tension between them being unresolved in Kuhn's published works.[2] Kuhn might not have acknowledged the divergences between these three strands but it is perhaps his failure to do so that gave him the most trouble and left him unsatisfied.

We have largely kept our own – Wittgensteinian – views in check, though it may have been apparent that we find Kuhn most defensible when closest to offering *therapeutic* possibilities, when dissolving the philosophy of science, or at least subverting and overcoming the main traditions (the 'received views') in philosophy of science. Our bringing out of this neglected therapeutic aspect amounts to a *recovery* of Kuhn's thought, making that thought available to a contemporary intellectual scene that falsely considers that it understands and has absorbed (or refuted) Kuhn. His 'Wittgensteinianism' is present in his 'underlabouring' for the historiography of science, in his 'anthropological' attitude to past science; and crucially in his distrust of normative ambitions in the philosophy of science, and his scepticism of explanations of 'the real nature of science' or of 'how scientific progress is possible'. It is crucial also for his emphasis on science 'as nitty-gritty' (thus distinguishing science properly from metaphysics, that is, from speculation); and part and parcel of his displacement of epistemology by description, and his challenge to the fantasy of a transcendental standpoint in philosophy.

However, we do not want to – and could not possibly – claim Kuhn as a thorough Wittgensteinian; for Kuhn wants to be a good philosopher too, and for him that means in the analytic tradition represented by Quine, Davidson and Putnam.[3] In his skirmishes with these thinkers Kuhn more than holds is own, and maintains that his notions of incommensurability and world changes do not entail the errors that they are criticized for. Even so, Kuhn is too ready to accept that he needs to elaborate a philosophical rationale for his notions and is drawn into treating *metaphysical* issues more 'seriously' than he need. In doing so, he neglects other of his own – 'therapeutic' – arguments. Kuhn's later efforts to make his meaning more precise, to defend ways of speaking he had initially adopted provisionally, often tend to reification, or a theorization, to an extent greater than we accept. Sometimes Kuhn's tendency to metaphysics and to 'theorizing' (as in his later hopes of contributing to a new 'theory of meaning') tends somewhat towards a 'relativistic' metaphysics (for example, on occasion in relation to 'world changes', as discussed in chapter 5). Sometimes the covert metaphysics is more empiricism or positivism – we have discussed this in relation to his failure fully to transcend the myth of 'the given'. Hoyningen-Huene's invaluable 'reconstruction' of Kuhn's thought plays into Kuhn's tendency to a 'relativistic' (pluralized 'Kantian') metaphysical view, and fails to recognize the 'Wittgensteinian' aspect of Kuhn's work.[4] Recent *critics* have tended instead to emphasize the positivistic philosophy latent in some moments of Kuhn's work.[5]

The two most recent book-length studies of Kuhn – Bird's and Fuller's – have been from completely different standpoints: Bird's a naturalistically inclined analytic philosophy; Fuller's from politically oriented Science Studies. Both miss Kuhn's therapeutic moment almost completely, emphasizing the positivist affinities.[6]

There is also a *naturalistic* streak that appears overtly on odd (very odd) occasions. Bird believes there is not enough of this in Kuhn. We think the reverse. The major manifestation of Kuhn's naturalism (cf. *RSS*, 95) is his enthusiasm for Darwinism (and for quasibiological renditions of the nature of science), and for outlining thereby what seems to be a fully serious intention to consider the development of scientific specialisms in terms of natural selection, with 'incommensurability' as a mechanism of speciation, insulating different specialisms. We cannot see how Kuhn could have seriously pursued this idea further, nor why he should think it would gain anything for him (the idea of the 'evolutionary niche' could serve his philosophical purposes on a metaphorical basis). An earlier outburst of such gratuitous naturalism arose in his 'Second thoughts on paradigms' when he abruptly announced that his account of the way a child learns to discriminate between water fowl is a process that

> can easily be modelled on a computer; I am in the early stages of such an experiment myself. A stimulus, in the form of a string of n ordered digits, is fed to the machine. There it is transformed to a datum by the application of a preselected transformation to each of the n digits, a different transformation being applied to each of the n digits, a different transformation being applied to each position in the string. Each datum thus obtained is a string of n numbers, a position in what I shall call an n-dimensional quality space. In this space the distance between two data, measured with a euclidian or a suitable noneuclidian metric, represents their similarity [etc., etc.]. (*ET*, 310)

No more is ever heard of it – thankfully, in our view. But the naturalistic strand is throughout. We are not sure how deep this strand runs, but suspect it is one implied in Kuhn's self-reflexive tendency, that is, to think sometimes that his own efforts ought to be comprehensible in terms of his own scheme, leading him to think of what he is doing as close kin to science. This tendency to self-reflexive application is present both in *SSR* and periodically thereafter – Kuhn was indeed right to worry about self-refutation (his esteemed predecessors in Logical Positivism had self-refuted, their

Verificationist criterion ruling itself out of order) and to try to guard against this. His worry on this score had a seriously negative consequence. *It encouraged Kuhn to believe, erroneously, that what he was doing was actually itself a kind of science.* As Wittgenstein would put it, Kuhn has, like many distinguished thinkers before him, the alleged method of science before his eyes. One might say that this is not surprising, for an ex-scientist, for a historian and philosopher of science. One might also say, however, that it is especially ironic and unfortunate that, despite all Kuhn's achievements, he remained in too great thrall to the image of science as a *telos* for all human inquiry. Kuhn variously encouraged himself to think that his brand of philosophy was either superscience (metaphysics), or continuous with science (Positivism, or Quineanism), or simply science ('Naturalism'). This provided a further bulwark against the likes of Barnes, Fuller, Sandra Harding, Barbara Herrnstein-Smith, Lyotard, et al.: if what Kuhn was himself doing was science, then it could hardly be the 'soft' political/sociological/historical stuff.

Kuhn played, dangerously, with the idea that the philosophy of science was in crisis, that he himself would help to engineer the new paradigm (*SSR*, 77–8). This involved forgetting the deflationary/ therapeutic dimensions (and also the historiographical ones) of his own practice; it involved dangerously prognosticating the future of philosophy of science; it involved failing to hold clearly in view that the philosophy of science is philosophy, not science. It involved, in short, a lapse into scientism.

The Positivists were right to be dismayed that their practice failed their own criteria, because for them philosophy was through and through (continuous with) science.[7] But *Kuhn* had the resources available not to make that mistake – he didn't *need* to *worry* that what he was doing could not meet the 'criteria' (that is, the existence of paradigms, etc.) he found exemplified in real science (as opposed to programmatic – pseudo – science). He should have been content to be a philosopher (and a historian etc.). But he couldn't help wanting the prestige of science, and the 'protection' against misunderstanding apparently involved in the status of (postpositivist) empiricist philosopher, or of (relativistic) metaphysician, or of 'scientific naturalist' (psychologist, biologist, computer programmer). This adds nothing to and engenders confusion about the real substance of Kuhn's achievement.

Where Kuhn is at his best – in the deflationary aspect of his thought – is in the following important suggestions on how to think about science:

- That science is about the detailed understanding of phenomena.
- That paradigm innovation is rooted in knowledge of the phenomena and not in any philosophical understanding of truth (let alone of 'methodology').
- That the unit of science is the solved problem, and that scientific innovations gain their purchase by virtue of 'technical bite'.
- That revolutionary paradigms are originally unfinished productions.

These are not the attention-grabbing formulations that attracted fame and misfortune to Kuhn, but they are the ones that make clear, first of all, that he does not view science at all as philosophers of science are apt to do, which is as the fulfilment of metaphysics, yielding the *ultimate* categories for the characterization of nature (which is assigned, by those metaphysically inclined, the status of ultimate reality). There is no denial that scientific categories 'fit nature', but they do so *in a way, to an extent, relevantly to certain problems and relative to a paradigm*, and perhaps most important, in a way some relevant scientific community deems better than previous paradigms. It is not the business of science to produce a set of general categories of the kind that metaphysics seeks. Kuhn has done away with the idea of the philosopher of science as the transcendental adjudicator of different scientific schemes (though his Darwinian 'naturalism' is a covert reintroduction of a transcendental standpoint), who determines which of rival paradigms does 'best fit' nature, for that judgement is one which is *unavoidably* internal to science itself, where the question of 'fit with nature' takes the form of highly specific configurations of intensely technical issues, and the decision between paradigms (in so far as it is a decision at all) is an assessment as to the success and prospects of a solved problem.

Our first basis for appraising Kuhn in this book has been on the basis of his own arguments and relative to the objections of his critics. This is why we have placed such emphasis on trying to understand what Kuhn does say. If one does that, then one sees that his case is carefully argued compared with that of many critics. It is not always the case that Kuhn's ideas are well argued, and it is sometimes plain that Kuhn is in serious difficulties (frequently of his own making). It is plain that on occasion Kuhn has got himself into an uncomfortable position and is thrashing around. Only after we have made assessments of the stronger and weaker parts of Kuhn's corpus can we make the diagnosis we are making here: the more successful and least problematic parts are those tending towards the therapeutic, and the least satisfactory are those affiliating to the main traditions in Anglo-American philosophy (including 'scientific naturalism'). It does not matter whether Kuhn would or would not explicitly link his thought to Wittgenstein's (though of course Kuhn did himself on occasion[8]), for we can still find strong affinities.

Attempting to affiliate Kuhn and Wittgenstein in the way we are attempting here *may* do the former no favours, especially since there are so many who 'know' that the therapeutic view of philosophy *can't* be right, and will respond that Kuhn the wild idealist, relativist and irrealist – while wrong-headed (or never actually having existed) – just is so much more interesting than Kuhn the 'quietist'.[9] But Kuhn's 'Wittgensteinian' implications *are* dangerous, though in a way that the commentators are usually unwilling to recognize: they are dangerous to *the philosophy of science* rather than to science. The Wittgensteinian traits to be found in Kuhn diminish the content of the philosophy of science relative to what has traditionally been envisaged, and, if developed still further, would evaporate that pursuit altogether. As it is, the legendary Kuhn has had the greater influence, allowing commentators to continue with philosophy of science – even if having to change its name to (say) 'sociology of scientific knowledge'. Their creation of the legendary Kuhn has been used to license continuation of the argument over whether science 'represents reality' or not. In Wittgensteinian terms, argument over this global question is futile, since it is only (in Kuhn's terms) in the context of a scientific controversy itself, and as part of that controversy, that the question as to whether this or that paradigm best represents reality can be raised, if at all – and the question is to be answered by further *scientific* controversy and further *scientific work*. The hard lesson of Kuhn (as of Wittgenstein) is that the questions as to whether science represents reality or makes progress *are not philosophical questions about science*, but can only ever be misleadingly abstracted versions of specific *scientific* questions.

Wittgenstein on this point is more resolute than Kuhn, who seeks to accommodate to Realism without giving up views that make him so uncomfortable for most Realists. Wittgenstein does not think anything that matters is lost if Realism is 'given up' since it lacks substantive content to differentiate it from supposed rivals such as Phenomenalism or Idealism: the disputes 'between Idealists, Solipsists and Realists look like [this]. The one party attack [*sic*] the normal form of expression as if they were attacking a statement; the others defend it, as if they were stating facts recognised by every reasonable human being.'[10] That all the sound and fury signifies nothing is perhaps too awful for many to contemplate.

Kuhn is like a Wittgensteinian dissolutionist and much less the 'Realist' in thinking that *'Realism' does not matter to the motivation and progress of science*. Kuhn too shows that the central problem of the philosophy of science – about the relationship between scientific schemes and the intrinsic properties of nature – is wrongly posed. However, there is no further court of appeal – the only possible talk about the intrinsic properties of nature must take place in and through the paradigm-based work of scientists. There is no way of standing back from the succession of paradigms and asking for an independent way of determining whether this is a progressive movement, whether there is a continuing advance towards determining the intrinsic properties of reality itself.

It may seem like cynicism to hold, with Kuhn, that the succession of paradigms *is* progress, since the story of scientific change is characteristically written by the victors – a version of might-makes-right perhaps? But Kuhn's picture is *much less voluntaristic* than many 'constructivists' want it to be – the respects in which for Kuhn science resembles the arts are palpably not respects licensing an 'anything goes' approach (even in the arts). This is why, contrary to Feyerabend's allegations, Kuhn's work cannot be understood as involving, even ambiguously, a methodological injunction of the sort Feyerabend thinks he sees – the ways in which not just anything goes are *built into* the *nature* of the discipline(s), and neither require nor can be given 'supplementation' by methodologists/philosophers. Postmodernists, 'social epistemologists/constructivists', etc., tend to think that Kuhn has licensed voluntarism in science – this is a complete misunderstanding. Even philosophical Realists are not as clear as Kuhn is on the ways the concrete reality of scientific practice makes their dreams of rational choice between theories hopelessly abstract! Kuhn is actually the least voluntaristic philosopher of science there is!

Kuhn's central 'argument' is *descriptive*. As Kuhn's implications for the social sciences showed, Kuhn's 'argument' is: *this is how things have happened in the natural sciences*. Paradigms *have* emerged, there *have been* revolutions. The development of science *has narrowly constrained* the range of possible options, even at points of crisis. The idea of individuals attempting to foment scientific revolutions or counterrevolutions for their own sake is just meaningless in that context. In science unified around a paradigm scientists will be working on their problems, and they *won't be interested* (let alone have time to be) in stuff that does not bear upon their problems. That someone else is attempting to develop a new paradigm when, so far as they can see, there is no *need* for one is hardly likely to seem other than perverse. Feyerabend wants scientists to be keen on the proliferation of paradigms *on (democratic and epistemological) principle* – either this is what scientists would do, if the matter was made clear to them, or, if they do not, then they ought to be converted into the kinds of people who *do* value these principles. Kuhn's point is that the unification around a paradigm is not a principled matter in that way at all. People subscribe to the paradigm simply because they subscribe to the paradigm: it *speaks to them* in ways that other scientific doctrines don't, they are convinced by it even if others aren't and are *categorical* in their attachment to it. Attempting to make changes in science of the sort that Feyerabend envisages would not just be a matter of introducing something new – deliberately proliferating paradigms – into a science that could otherwise remain unchanged, but would, if Kuhn is accepted, be a matter of having to change pretty much everything about science, including the kind of people scientists are.

One of the things that underlies the typical philosophical concern with rationality (and much thinking about people's activities in the social studies) is that doing things involves deliberative decisions, people weighing up pros and cons, and opting for the most favourable balance of pros and cons – so-called models of rational action. But, as Kuhn emphasized throughout, the way that scientific development *pans out* does not originate in deliberative decisions: the extent to which people react and respond to circumstances is much more important – they characteristically do not *choose* to respond as they do, they 'just' respond as they do. Thus, it is just that people *can't find something interesting*, that they can't get the hang of it, that it leaves them cold, that they can't leave it alone and must keep picking at it, and so on.

Many of us still need to be reminded that there just is no way to nature other than through scientific paradigms, and therefore through their invidious comparison. There is no such thing as comparing paradigms as live alternatives other than through engagement in the revolutionary controversy, and no live-option basis for comparing them other than those that scientists themselves use. The considerations involved are characteristically highly technical and markedly prospective, not to mention relative to concerns and sensibilities. As the sides line up during a revolutionary crisis, there are not perhaps any categorical considerations which favour one side over the other, and the pros and cons bandied about can affect any individual distinctively, while the future prospects of the paradigms is a matter of mere (if educated) speculation. Anyone who wants to adjudicate between live paradigms is taking part in – not standing aside from – the struggle, and only someone with knowledge of and a stake in the issues can *seriously* take part. When the revolution is over and the shift has taken place then only one paradigm is any longer a live option. Exercises in showing that the science might have persisted with a rejected paradigm rather than gone over to the new one (*cf.* Pickering[11]) are of historical interest only. The whole logic of *SSR* is that sciences never reinstate rejected paradigms (as such). The scientific revolution, when it is over, is water under the bridge. That there is nothing at the time to vindicate a paradigm shift does not mean that subsequent developments cannot provide vindication (nor do we need to think that the shift *must* somehow, sooner or later, be vindicated – that it has been made is the fateful thing). The nature of scientific revolutions is such that within the sciences themselves there is no one except 'the victors' left to write the story.

A caution is called for. We must be very careful not to suggest that Kuhn is – in his histories – trying to do something that we are now claiming his writings about scientific revolutions say cannot be done. Kuhn's histories are meant to provide – we use the words without apology – *objective historical accounts* of what went on at a particular time in a scientific revolution, an aim which he counterposes against that of writing 'Whig history', history 'from the victors' point of view'. Any contradiction is only apparent, for the story which Kuhn seeks to tell is of a different kind from that the scientific 'victors' will tell. Kuhn's accounts have no stake in the scientific outcome of paradigm changes, while this is the very thing that matters to scientists for whom the paradigm now in place is the measure of scientific achievements. The histories Kuhn can write will not give scientists reasons to rethink their contemporary science – they cannot scientifically reanimate the past. The reasons why scientists accept the paradigm they do are (intrinsically) comparative ones and they provide (in the science) the reasons why it is better. There is nothing else for them to say. Their predecessors' paradigms are no longer scientifically credible, nor does Kuhn seek to rescue them from rejection, only from the condescension that sometimes accompanies rejection. Neither the historian nor the philosopher can second-guess the scientist.

An evil influence?

Arguably, Kuhn (although not a 'proper' philosopher) has been not just the most influential philosopher of science in the second half of the twentieth century, but the most influential philosopher, full stop, and tended to be at his worst when *trying* to be one of those. Kuhn's position with respect to science is really, as we have already suggested, parallel with Wittgenstein's in respect of philosophy: leaving everything as it is. It might be thought that Kuhn's influence is testimonial to the falsity of Wittgenstein's assertion, for Kuhn has had, willy-nilly, a tremendous influence and, intentionally or not, has contributed to a transfiguration of the intellectual landscape. Kuhn has been recruited to the 'science wars', and the 'culture wars' more generally. He is subject to vitriolic assaults precisely because he is a totemic figure for the 'irrationalist' forces marshalled against the realist and rationalist forces siding with science. Kuhn is invoked by both sides in these disputes as a major – if only one of several – sources of 'relativism' in contemporary culture. Kuhn certainly has not 'left everything as it is', it might be said.

One thing wrong with 'realism' is that it invites 'relativism': the dissatisfactions it gives rise to encourage the formation of opposition, and one form this takes is the 'relativist' mode. But if 'realism' is without substance, then so too must 'relativism' be. The implication: that 'relativists' are not – except in their inclination to recite supposed relativist doctrines – relativists at all. Wittgenstein's contention that philosophy leaves everything as it is can be recast as the claim that philosophical opposites nonetheless agree with one another in their unreflective practice. Philosophy's leaving everything as it is just does not preclude it from altering philosophical opinions, but refers only to the fact that it need not and generally does not thereby modulate pre-existing agreement in practice.

Who has ever really supposed that just anyone can enter into an area of scientific work and make an effective intervention? Who has suggested that any innovatory contribution has nothing whatever to do with the previously extant results of science? Who imagines that anyone can – and arbitrarily – offer a new scientific paradigm and anything one can cook up, that anyone can just *decide* that's how nature is? Everything that Kuhn says amounts to: natural science is *anything but easy*.

Of course, philosophy of science does make differences: we'd like our *philosophical* intervention to help end a long and utterly futile dispute, we'd like people to see that the 'science' and 'culture' wars have been between largely naked emperors, and that what people thought were reasons to devalue science or to take up arms against Realists are only *bad* reasons – as bad as the reasons for sticking with the Realists. But even if our intervention were successful, dispelling a great cloud of *philosophical* confusion, our philosophy does not entail direct ethical or political consequences. There are questions to be asked about the role of science in our society, the responsibility of scientists, etc., but our philosophy will not give answers to these questions. Nor can it, but then neither can *any other* philosophy honestly, responsibly and correctly answer these questions – *for they are not philosophical questions*. The fact that *philosophers* cannot tell you how to live should not stop anyone making up their minds on the issues of the day – after all, in the contemporary cultural climate, philosophies are often chosen because they seem to be a crutch, a further support, for convictions you already hold. We are comfortable to hold our own convictions with whatever confidence – even certainty – we have in them without needing to garnish them with a further conviction: that what we think is not just right as we see it, but demonstrably in accord with the intrinsic nature of things.

To conclude

It is an irony that we have sought to optimize the consistency of Kuhn's thought in a way which, if accepted, would clearly deprive him of many of the characteristics that have won his reputation and earned him his influence. Thomas Kuhn's account of science definitely does not give most philosophers of science what they (think they) want from a philosophy of science. Kuhn, we have argued, neither provides a general and true theory of science, nor a set of normative prescriptions for how to pursue science correctly. That Kuhn does neither of these can seem hugely disappointing, which is perhaps why friends and foes alike seek to read a theory and/or some prescriptions into his work. However, 'deflating' Kuhn is not meant to diminish but to fortify his arguments. Kuhn's strength is his 'negative' thinking. Kuhn tries to get one to rethink what the philosophy of science is, not to manoeuvre within traditions of that discipline. One might even say that he attempts to cure us *of* the philosophy of science. It is this which makes him truly radical. Indeed – revolutionary.

Why do we praise Kuhn for 'negative' thinking? Lack of understanding is not necesssarily the same as ignorance, but it is often the result of a lack of clarity in one's thought, and the lack of clarity can result from difficulties one puts in one's own way. One of the illusions which often has a strong hold on the mind is that the way to a better understanding is through the addition of further knowledge, but this is not always the case. Sometimes, the difficulty is not that we don't know enough, but that we don't properly appreciate what we already understand. What is then needed is to get a more perspicuous grasp on things we already know, perhaps by rearranging the familiar facts. We have written here very much in the spirit of the following remark of Ludwig Wittgenstein's:

> Philosophy may in no way interfere with the actual use of
>   language; it can in the end only describe it.
> For it cannot give it any foundation either.
> It leaves everything as it is.
> It also leaves mathematics as it is.[12]

We would add that it leaves science as it is too, and that Kuhn has (largely) shown us how to see this. *That* is a (different kind of) 'revolutionary' achievement.

Notes

Introduction: The Legendary Thomas Kuhn

1 Steve Weinberg, 'The revolution that failed', *New York Review of Books*, 8 Oct. 1998, pp. 48–52.
2 Robinson, *Philosophy and Mystification*, p. 192. Robinson's treatment of Kuhn's work is unusually fair and empathetic.
3 Weinberg, 'The revolution that failed', p. 48.
4 Ibid.
5 For Kuhn, thinking of science as practice and as work and as making as well as finding is often therapeutic, and one might then try the following analogy, to technology: for, after all, though there clearly has been progress in automobile design, nobody says that we are progressing towards the true car, or even towards the true understanding of how to makes cars!
6 Weinberg, 'The revolution that failed', p. 52.
7 Ibid., p. 50.
8 Closer analogues are perhaps the following: the legendary Peter Winch, Relativist star/bogeyman in the philosophy of the social sciences; the legendary Wittgenstein himself; and also the legendary Karl Marx, whose own attitude to his 'followers' was intriguingly similar to Kuhn's.
9 See pp. 263–4 of Kuhn's 'Reflections on my critics' in Lakatos and Musgrave, *Criticism and the Growth of Knowledge*. Care is needed in interpreting Kuhn here; this is not a blanket castigation of Feyerabend but is primarily criticizing Feyerabend's rhetoric. Kuhn believes that Feyerabend is pretty close to Kuhn, and that Feyerabend's rhetoric distracted even more than Kuhn's own way of writing distracted. For discussion, see also Richard Bernstein's excellent *Beyond Objectivism and Relativism*, p. 59.

10 i.e. to hasten the end of history of science as merely a popular pursuit/an activity of (ex-)scientists in their dotage/a mode of legitimation of current science and little more. For detail, see Hoyningen-Huene, 'The interrelations between the philosophy, history and sociology of science in Thomas Kuhn's theory of scientific development', pp. 488ff.

11 For discussion of this theme in relation to Kuhn, see Richard Rorty, 'Being that can be understood is language'. Weinberg seems to be indignantly affronted by the way Kuhn degrades the image of science; Rorty's take is different.

12 Suppe, *The Structure of Scientific Theories*.

13 Hacking, *Representing and Intervening*.

14 Ibid., pp. 5–6.

15 Ibid., p. 6.

16 See our discussion, 'Incommensurability 1', chapter 4, below; and *SSR*, 98–102.

17 In what sense, Kuhn himself discusses in the final chapter of *SSR*.

18 This is the doctrine of 'Metaphysical Realism', the notion that the Universe has, as it were, a preferred way of being described.

19 One account which refrains from blaming Kuhn for anything is Hacking's *Representing and Intervening*, pp. 2ff. A clear example of the opposite tendency is Lakatos.

20 See, for instance, Hanson, *Patterns of Discovery*.

21 Specifically, as we will see below, he argues that at best you can have an observation language *within a paradigm*; see *SSR*, 129.

22 Not a theory, nor even, strictly, a model. Just a sketch, a picture, for certain 'practical' and 'therapeutic' purposes. (*Contra*, for one example of many, Fuller, *Thomas Kuhn*, who endlessly presupposes that Kuhn's central thrust is to produce a theoretical model of science.)

23 Hacking, *Representing and Intervening*, p. 8.

24 Ibid., p. 9.

25 Though this is not to say that there is *univocal* progress through revolutions – see below.

1   *The Structure of Scientific Revolutions*

1 As Hoyningen-Huene points out in *Reconstructing Scientific Revolutions*: 'In fact, despite proposing henceforth to make this substitution of "disciplinary matrix" for one sense of paradigm, Kuhn does not in fact make subsequent use of either of these formulations' (p. 143). The basic reason for this, we take it, is that he just ceases to discuss paradigm in this sense very much at all – *because it was never of primary interest to him*. For him, a philosopher of science, and sometimes a historian looking at examples, but rarely a social scientist building up data about the actual institutions of science, the more important aspect

was paradigms as ways of acting/practising by means of exemplars. (A vital contrast with Fleck, which makes Fleck far more suitable for use in 'Science Studies', 'Sociology of Scientific Knowledge', etc., is, roughly (because slightly anachronistically) speaking, that Fleck, unlike Kuhn, treats disciplinary matrix and not exemplar as paramount – and thus allows that *any* discipline can have a paradigm. Cf. Rouse, *Knowledge and Power*, pp. 30f.)

2 We find it convenient to use the famous and evocative word, 'paradigm', and are not as ready as Kuhn was to give up the term. To anticipate slightly: Generally, what we mean by the word, and this is very important, is a disciplinary matrix *constructed around an exemplar(s)*. This combines together but in a particular and a reasoned way – i.e. a way 'demarcating' (the distinctiveness of) real sciences – the two main senses of the word 'paradigm' distinguished by Kuhn in the Postscript to *SSR*, discussed below. For support for this version of how to understand 'paradigm', see Buchwald and Smith, 'Thomas S. Kuhn, 1922–1996': '[Kuhn held] that scientists learn to practice their disciplines through exemplars, and that in the process they assimilate a way of thinking that forms the core of a disciplinary matrix' (p. 367).

3 Presumably, their work was of a suitably serious and empirically oriented nature to justify such a designation. Even so, it does not follow that 'social scientists' past or present are therefore also *automatically* entitled to consider themselves scientists, in a comparable position to the forerunners of physical theories of light. See chapter 3 below.

4 Compare Charles Darwin (quoted in *SSR*, 151): 'I by no means expect to convince experienced naturalists whose minds are stocked with a multitude of facts all viewed, during a long course of years, from a point of view directly opposite to mine . . . But I look with confidence to the future – to young and rising naturalists . . .'

5 Ignoring for present purposes the fact that the term had been used in the philosophy of science prior to Kuhn, at least in the crucial sense, roughly, of 'exemplar'; see Hoyningen-Huene, *Reconstructing Scientific Revolutions*, p. 132.

6 See her essay, 'The nature of a paradigm', in Lakatos and Musgrave, *Criticism and the Growth of Knowledge*.

7 Victorious paradigms are 'more successful than their competitors in solving a few problems that the group of practitioners has come to recognise as acute' (*SSR*, 23).

8 On the rare occasions when this does not happen, Kuhn writes (*SSR*, 79), the area of science in question ceases to be so, and becomes simply a branch of engineering.

9 See also Fleck, *Genesis and Development of a Scientific Fact*, pp. 85–6.

10 There is a danger in assuming that shared modes of expression do not conceal substantive differences among scientific contemporaries, in the same way that they may exist between precursors and successors, even if they use the same terms. It may turn out at a time of crisis *either*

that in a sense there had never really been full agreement on the paradigm, *or* that the full agreement that perhaps had actually existed nevertheless disintegrated when confronted with an unprecedented and distinguishing case. Cf. Hacking, 'Rules, scepticism, proof, Wittgenstein'.

11 Note Kuhn's striking claim that 'Normal science does not aim at novelties of fact or theory and, when successful, finds none' (*SSR*, 52).

12 One can also see the difficulty – in one important way insuperable? – in attempting to speak entirely without 'presentism'. For the description of the compound in question as 'red oxide' would not have been available prior to the Lavoisierian chemical revolution. *Kuhn's* language is sometimes hesitant, even 'tortured', because of his scrupulous efforts to speak honestly and non-Whiggishly about past scientists' work.

13 One ought to think of 'times' of revolution as being characteristically much longer than is perhaps intuitive or traditional. Kuhn brings this out not only in *SSR* but in detail in *CR* (and *BB*). It later leads him to suggest that it may be better to think of scientific revolutions as being sometimes 'experienced' *only by historians*, not by individual scientists.

14 To say that this is an innovation of Kuhn's may seem misleading: didn't the Positivists emphasize questions of sense (in setting out the groundwork for any regime of confirmation and information)? Yes – but not until late Carnap did they do so in a relatively sophisticated manner. There is a greater degree of anticipation of Kuhn in mature Logical Empiricism (especially in late Carnap) than has (until recently) been noted, even by Kuhn; but this is because in late Carnap and Hempel one has a view *which has already moved some considerable distance in Kuhn's direction*; and indeed, these figures (especially Hempel) came to be *influenced* by Kuhn himself. Cf. *RSS*, 226f.

15 See the Kuhn's essay, 'Objectivity, value judgement and theory-choice' in *ET*.

16 This is a remarkable, almost Rortyan, admission on Kuhn's part. One way of making sense of the point – a way highly consonant with Hoyningen-Huene – would be to express it as follows (emphasis ours): '[T]hough the *noumenal* world does not change with a change of paradigm, the scientist afterward works in a different *phenomenal* world.' Even so, Kuhn is, we believe, more aware (than Hoyningen-Huene tends to suggest) that a pluralized Kantianism has its own problems (principally, semantic relativism).

17 Part of the difficulty here – and a reason Kuhn is often (mis-)characterized as an irrationalist – may be that words like 'observe' and 'perceive' are frequently employed as 'success words'; i.e to use the word 'observe' is often to imply directly *that what was observed is really there.* So, given that Kuhn does not seem to endorse theory-neutral observation, people think that what he must be saying is that what is

observed is therefore really there . . . We are intimating that, instead, Kuhn's cautious talk about the importance of changes of vocabulary should be read as about *the historian's task*, not about metaphysics or relativism. Kuhn is talking about understanding people who would say without any trace of irony or distance that (e.g.) 'I observed the moon and others of the planets last night.'

18 Again, this unease alone should be enough to impress upon us that he is not doing what his opponents have usually thought he is – namely, firing off outlandish ('Relativist', 'Idealist' or 'Irrationalist') philosophical theses for our delectation or horror. It is crucial to bear in mind that *every single time* that Kuhn in *SSR* uses one of the expressions which trigger accusations of 'Idealism', etc., he qualifies it, as something he is e.g. 'inclined' to say. He is well aware that, to adherents of the traditional 'epistemological paradigm' in philosophy, he will sound as if he is then speaking nonsense, but *that* is one reason to give up the traditional paradigm – it makes the proper writing of history of science problematical.

19 Cf. Baumberger, 'No Kuhnian revolutions in economics'.

20 For further discussion see *RSS*, 161.

21 Sankey, *The Incommensurability Thesis*, p. 161.

22 Kuhn's attempts to sort out whether paradigm shifts are to be understood perceptually, psychologically, linguistically, or what-have-you – half-forced on him by his critics in the decades following *SSR* – is a sign of not having fully accepted that there is no philosophical theory to be had here, and no need for one.

23 Are we committing ourselves to some dubious version of the analytic versus synthetic distinction here? No, only to a picture of our language suggested by Wittgenstein's *On Certainty*, paras 95–110 (cf. *SSR*, 141), and we are resisting only an extreme Quineanism that would rule out *any* differentiation, even a subtle one, between facts and concepts.

Are we disagreeing with Kuhn, in that he sometimes speaks as if in fact it is all theories, not just paradigm-central theories, which are at issue (see e.g. *SSR*, 80 and 96)? Kuhn tends to use the word 'theory' where we would be inclined to use the expression (e.g.) 'paradigmatic theory', because he tends to class only those scientific innovations which shake up their field in a surprising way as 'theories'. Otherwise, he describes what is happening as only theory *articulation*. See *SSR*, 97: 'research aims at the articulation of existing paradigms rather than at the invention of new ones. Only when these attempts at articulation fail do scientists encounter . . . the recognized anomalies whose characteristic feature is their stubborn refusal to be assimilated to existing paradigms. *This type alone gives rise to new theories*' (emphasis added).

24 This should not be understood to imply a theory of language, e.g. a quasi-Carnapian 'grammar theory'.

25 See Wittgenstein *Philosophical Investigations*, para. 7.

26 See again Rouse, *Knowledge and Power*, and Hacking, *Representing and Intervening*, pp. 62f.; Hacking's book foregrounds very strongly the practical non-linguistic activities of scientists, following Kuhn (e.g. *SSR*, 142).
27 Cf. also Kuhn's explicit remark on *SSR*, 160 that the 'issue' of progress in science is to a reasonably substantial degree semantic in nature.
28 Conant and Haugeland's formulation in their Editors' Introduction to Kuhn's *RSS* is useful here: '[Scientific] progress takes the form of ever-improving technical puzzle-solving ability, operating under strict – though not always tradition-bound – standards of success or failure' (2).
29 And even this gives one few or no grounds for Lakatosian prognostication (i.e. normative philosophy of science) as to the (proper) future direction of the science.

2 The Historical Case Studies

1 With the use of 'sphere' now not referring to spherical shape, but putting the emphasis on separation and difference.
2 Cf. Feyerabend, *Science in a Free Society*.
3 Asimov, *Understanding Physics*, vol. 2. p. 139.
4 Kuhn and Heilbron, 'The genesis of the Bohr atom'.
5 Ibid., p. 212.
6 Ibid., p. 276.
7 A preliminary characterization of the general relations between Kuhn and his main predecessors in philosophy of science: Kuhn does not disagree with the Logical Positivists and with most varieties of Empiricists on the basic notion of accumulation of scientific knowledge – except he restricts it to periods of normal science. He does not disagree with Falsificationists over the idea that major theories can come to an end and scientists have to try out a new hypothesis – but that only happens very rarely. He rejects the (Falsificationist) thought that scientists standardly look for major innovations, and the (Positivist) thought that such innovations can absorb the theories preceeding them. He rejects the 'formalism' – the fantasy of a logic of scientific method – that Positivists and Falsificationists alike have often shared. That means a rejection, above all, of any clear distinction between context of justification and context of discovery. And with that disappears the extremely widespread fantasy of a normative philosophy of science.

3 Kuhn and the Methodologists of Science

1 Lakatos and Musgrave, *Criticism and the Growth of Knowledge*.
2 Ibid., pp. 197–200. Cf. also Feyerabend, 'Two letters'; Preston, *Feyerabend*, p. 15.

3   Feyerabend, 'Two letters', pp. 354–5.
4   Feyerabend, 'Explanation, reduction and empiricism', p. 60. Feyer-
    abend will later change this 'view' markedly (cf. the early chapters of
    his *Against Method*).
5   The reader might sense in this sentence an implication that Feyer-
    abend may usefully be regarded as an unorthodox Popperian, but still
    a Popperian. It might seem remarkable to call Feyerabend, who
    scorned Popperiansm in his later works, a 'Popperian' but evidence
    marshalled in Preston, *Feyerabend*, suggests that the term would apply
    until as late as 1970, at least.
6   Barnes, *T. S. Kuhn and Social Science*, pp. 55–7.
7   Ibid., p. 46. See also Latour and Woolgar's *Laboratory Life*, implying
    that a Nobel-prize-winning discovery can now be achieved by a team
    of scientists *entirely* engaged in 'mundane' research.
8   *SSR*, 23; though cf. note 4 on *ET* 295, for a somewhat different
    emphasis.
9   Kuhn came to think it was about a hundred people who typically
    worked in the shadow of an (exemplar type) paradigm.
10 And, again, where are the critics on this point?
11 i.e. unlike the philosophers of science whose abstract methods Kuhn
    wishes to supersede, he claims to be about to *exemplify* his points *prior*
    to theoretically making them: e.g. 'abstract discussion will depend,
    however, upon a previous exposure to examples of normal science or
    of paradigms in operation' (*SSR*, 11). Our point is that he actually falls
    short of delivering on the latter claim.
12 In a *different* sense, anticipated in chapter 1 above, and drawn out
    further below, they do reveal a certain 'radicalism': in the idea that sci-
    entists nearly *always* work normally, even when the results are extra-
    ordinary. This therefore is in fact one important element of the answer
    to our question: Kuhn's account of normal science is less unevidenced
    than it might seem *because his concrete accounts of scientific revolutions
    are themselves in one very central respect accounts of normal science*.
13 The 'revolutionaries' sometimes have be dragged kicking and scream-
    ing into 'their' revolution, cf. *RSS*, 105–20.
14 Lakatos and Musgrave, *Criticism and the Growth of Knowledge*, p. 237.
15 Ibid., pp. 52–3.
16 Contrast Fuller, *Thomas Kuhn*, pp. 74, 172–5.
17 Of course, it is conceivable that there *might* be a connection: for
    example, increasing corporate organization of science conceivably
    could sometimes delay normal scientific activity in some areas (and
    perhaps speed it up in others). Kuhn was happy to allow for tempo-
    ral effects of 'external' factors; and somewhat concerned by possible
    deleterious effects on the speed of growth of knowledge from the
    increased dominance of 'big [corporate etc.] science'. But these points
    make *no difference whatever* to his fundamental points about the phi-
    losophy of science (again, *contra* Fuller).

18 Pickering, *Constructing Quarks*.
19 And this has to do with the importance in Kuhn's thought of 'the essential tension' between tradition and innovation, cf. *ET*, 225–39.
20 Cf. Truesdell, 'Reactions of late Baroque mechanics to success, conjecture, error and failure in Newton's *Principia*', and Duhem's discussion of Wiener in *Aims and Structure of Physical Theory*.
21 One regrettable consequence of scientists' perhaps inevitable lack of detailed knowledge about earlier work is that a certain amount of retreading and repetition will be the present and future of large-scale science – Kuhn argues this at the end of his important 'The history of science', *ET*, 105–26.
22 Cf. Lakatos on 'sophisticated falsificationism' (*Collected Papers*, vol. 1, pp. 36–46); Popper's late statement on falsification (in *Realism and the Aims of Science*, pp. xix–xxxv), which moves to 'sophisticated' falsificationism, is among other things a major concession to Kuhn.
23 We can see now how badly off both Carnap and Popper, undoubtedly Kuhn's greatest predecessors in the philosophy of science, are, unless possibly our charitable rereading of Popper above can be fully worked out. Neither Carnap nor Popper actually succeed in accounting for even *part* of science!

Popper can't really understand even revolutionary science – because he tends to make it a matter of attitude rather than of *conceptual change*.

Carnap (a representative for us of the very best of Logical Empiricism – we are thinking of even fairly late Carnap, as understood according to recent revisionist scholarship) can't fully understand even normal science, because he doesn't really understand the room for manoeuvre and uncertainty already present there, let alone in revolutionary conditions: 'If positivistic restrictions on the range of a theory's legitimate applications are taken literally, the mechanism that tells the scientific community what problems may lead to fundamental change must cease to function. And when that occurs, the community will inevitably return to something much like its pre-paradigm state, a condition in which all members practice science but in which their gross product scarcely resembles science at all. Is it really any wonder that the price of significant scientific achievement is a commitment that runs the risk of being wrong?' (*SSR*, 101).
24 Cf. Kuhn's remarks on Sneed, *RSS*, 189–91.
25 Midgeley, *The Ethical Animal*, p. 51, final emphasis ours; nested quotes from Slavney and McHugh, *Psychiatric Polarities*, pp. 8, 123 respectively.
26 At least, such is *Kuhn's* important account of the failure of astrology to be a science, even in the days of its intellectual respectability, see Lakatos and Musgrave, *Criticism and the Growth of Knowledge*, pp. 7–11. We are serious when we say this is *one* possibility. There are others.
27 Feyerabend, 'Two letters', p. 355.

28 Feyerabend, 'Some observations on the decay of philosophy of science', pp. 24–5.
29 Winch, *The Idea of a Social Science*, pp. 86–7 (emphasis in the original).

4 Incommensurability 1: Relativism about Truth and Meaning

1   This is how Davidson, *Inquiries into Truth and Interpretation*, understands him (p. 184).
2   Namely, its denominator must both be an odd and an even number.
3   This way of putting the matter is closer to how Kuhn later put it.
4   Then again, we suppose Kuhn might argue that it is minute differences which make the difference between a theory being true and false, or between it being taken by scientists at the frontier of research to be saying one thing, or wholly another.
5   Or again, at least according to readings of Kuhn found in Davidson, Shapere, '*The Structure of Scientific Revolutions*', and Scheffler, *Science and Subjectivity*.
6   We will return to this in 'What is an ontology?', pp. 166–8 below.
7   Cf. Kuhn's critical take on 'referential essentialism', pp. 184–5 below.
8   Issues involved in this conception of translation are reviewed in Glock, 'The indispensability of translation'.
9   Hacking, *Why Does Language Matter to Philosophy?*, pp. 151–2. Hacking's account of Kuhn, and Kuhn's response to it, is in Horwich, *World Changes*.
10 See Donald Barry, *Forms of Life and Following Rules*, pp. 126ff.
11 Hacking, '*Why Does Language Matter to Philosophy?*, p. 151.
12 Do not be deceived by Davidson's use of 'radical interpretation', rather than 'radical translation'. Neither, for Kuhn, yet puts one in a position to translate. Understanding must come first .
13 However, it is important to Kuhn's later work that, in paradigmatic cases of scientific revolution, one cannot just add words (e.g. the old word 'phlogiston') into scientific language. Taxonomies of natural kinds cannot 'harmlessly' exist and overlap, in English. You cannot have 'oxygen' and 'phlogiston' both running around being *used* in sentences which are *used*. This goes even for historians, for the account they give will otherwise be simply incoherent.
14 'Dubbing and redubbing: the vulnerability of rigid designation', in *RSS*, 300. For more detail on 'Dubbing and redubbing', see chapter 5 below.
15 'Commensurability, comparability communicability', in *RSS*: 'I am concerned with a stronger version of untranslatability [stronger than mere unavailability of point-for-point translation], one in which not even longer strings are available' (note 11).

16 Mary Hesse, quoted by Kuhn in *RSS*, 54.
17 Winch, 'Understanding a primitive society'.
18 'Reflections on my critics', in Lakatos and Musgrave, *Criticism and the Growth of Knowledge*, p. 264. And see in *RSS*, 34: 'Most or all discussions of incommensurability have depended upon the literally correct [in Kuhn's terms] but regularly overinterpreted assumption that, if two theories are incommensurable, they must be stated in mutually untranslatable language'; and also *RSS*, 35, for Kuhn's motivated insistence that incommensurability does *not* imply incomparability.
19 'Reflections on my critics', p. 233.
20 Bernard Williams, 'Wittgenstein and Idealism'.
21 The most that is possible is for limited aspects of the old science, *considered sufficiently abstractly*, to 'return' in new colours – Kuhn argues precisely this case, against crude ideas of correspondence and of cumulation, in the case of Physics. But this is far from saying that we could undo, reverse or ignore the vast achievements and developments of modern science, and revert to the older science. There can be no going back to the game of Aristotelian physics, even if – in fact, precisely if – we come to understand that game. Whereas, arguably, it would be feasible to revert to certain central features of (say) Aristotelian Ethics – this is exactly what Virtue Ethicists are inviting us to do. They invite us to ignore or overcome peripheral aspects of that Ethics (e.g. the moments in it which appear to leave room for the political legitimacy of slavery), and abandon Kant and Mill in favour of believing and practising its central tenets. But there would *be* no centre to Aristotelian Physics without the irreversibly departed notions of (and practical procedures around) 'place' and 'motion' that Kuhn famously excavates for us.
22 Kuhn, 'Reflections on my critics', p. 264.
23 Wittgenstein, *Lectures and Conversations on Aesthetics, Psychology and Religious Belief* (emphasis added).
24 Why accent this word, 'ontology', a word little used by Kuhn himself? Answer: because it is a word much beloved by metaphysical philosophers.
25 Our argument is indebted to James Guetti's in *Wittgenstein and the Grammar of Literary Experience*. 'Grammatical effects' are the systematic effects that words have on one which are not well identified with their meaning, considered as communicative use.
26 Our argument here again puts Kuhn close to Hacking. For support of the view of incommensurability as centred on 'grammatical effects' (see *RSS*, 36).
27 Thought-style – 'denkstil' – was the very concept in Ludwig Fleck which partly anticipates Kuhn's 'paradigm'.
28 Wittgenstein, *Philosophical Investigations*, part II, p. 230 (emphasis added).
29 The scare quotes here are to mark again that this is *not Relativism*.

5   Incommensurability 2: World Changes

1   It is perhaps worth accenting that the 'world changes' idea is largely a *consequence* of incommensurability.

2   Alexander Bird (*Thomas Kuhn*, pp. 124–30), following Richard Bernstein (*Beyond Objectivism and Relativism*, p. 84), points out that Kuhn's ('dynamic') Kant is more like Hegel than Kant himself in two central ways (viz. antipathy towards noumena, and a 'dialectical' or at least dynamic approach to the growth of knowledge).

3   Having got himself into this position, Kuhn is as badly off as all 'post-Sellarsians', who, however much they try (cf. John McDowell's post-1994 work, e.g. 'Towards rehabilitating objectivity'), never quite arrive at Wittgensteinian therapy, which would dissolve their worries.

4   We think his move away from would-be 'cognitivist' and 'neuroscientific' approaches is a good thing but Bird (*Thomas Kuhn*, pp. 264–6, 280) sees matters differently, thinking that Kuhn should have become more scientifically naturalistic. See the Conclusion, below, for detail.

5   See Kripke, *Naming and Necessity*, and Putnam, *Philosophical Papers*, vol. 2: *Mind Language and Reality*, especially his 'The meaning of "meaning"'.

6   Bird, *Thomas Kuhn*, p. 183.

7   Kuhn wants us to understand that in an important sense scientists who are at all interested in the past have to regard their predecessors' whole schema and fundamental claims as wrong (*SSR*, 97); they cannot be contented, with positivists *or* essentialists, to regard them as only making narrowly specified factual errors. This is the intriguing and rarely recognized sense in which it is *Kuhn*, not his mainstream philosophical antagonists, who is properly prepared to have scientists (and informed laypeople) reasonably and unavoidably regard their predecessors as profoundly mistaken.

8   Bird, *Thomas Kuhn*, p. 183.

9   Kuhn in *RSS*, 102–3: '[the] niche to which they [creatures] are adapted is recognizable only in retrospect, with its population in place: it has no existence independent of the community which is adapted to it. What actually evolves, therefore, are creatures and niches together . . . Conceptually, the world is *our* representation of *our* niche . . .'

10   Kuhn argues that this so-called 'internal' world is not, in any case, the inner world of the individual thinker, but is rather the 'internal' world of the scientific community. The community is not possessed of a mind, and Kuhn thus gets himself off the charge (but only, and rather unconvincingly, on a technicality) that he regards reality as 'mind dependent'.

11   Garfinkel, *Studies in Ethnomethodology*, p. 76.

Conclusion: A Cure for the Philosophy of Science?

1　This way of putting things is suggested by Kuhn's collected and reflective suggestions on how to think of 'scientific revolutions' in *RSS*. It remains necessary to ask late Kuhn this question: if one thinks of 'scientific revolutions' like Kuhn does in *RSS*, as essentially a phenomenon only for historians, then what *exactly* is one to say of the situation faced in the (rare) truly extreme *and* more or less contemporaneous instances of conceptual change – such as the 'meeting' of Priestley and Lavoisier?

2　It is conceivable that Kuhn's *Nachlass*, the publication of what he was working on in the very last years of his life, may resolve the tension; but we think this very doubtful, for reasons to do with Kuhn's apparently deepening commitment to 'scientific naturalism' as his career progressed.

3　There is no reason, then, to think that Kuhn would embrace the 'programme' of determined dissolution of philosophical problems. There is clear evidence that Kuhn not only thought that philosophical theories could be had, but sought to construct them himself. When Hacking (in 'Working in a New World') suggested dispensing with the theory of meaning, Kuhn politely but firmly rejected the offer because a theory of meaning was essential to his conception of what he was doing (see *RSS*, 229).

4　A final note here on how, rather than through 'Kantian'/Darwinian spectacles, one might reformulate the 'many worlds' point: that scientists live 'in many worlds' is, after all, really only a striking way of putting the point that scientists work according to very different paradigms, and that their activities are therefore characterized by the – many – differences in what they do that follows from their accepting of this paradigm as opposed to that one. Kuhn's controversial formulations, in so far as they are acceptable at all, are only reformulations of other, unexceptionable observations.

5　It is important here not to be overly impressed or concerned by affinities between Carnap and Kuhn, in part because Carnap's view was (and became increasingly) sophisticated and in many ways unpositivistic.

6　Bird, *Thomas Kuhn*, Fuller, *Thomas Kuhn*. The picture in Fuller's case is actually very complicated: because at times Fuller writes as if it were not Positivism but a hidden political agenda motivating Kuhn. Yet Fuller also shows signs of recognizing that what he is really talking about is not Kuhn himself, but rather, 'Kuhnians'. Fuller's book sometimes works perfectly well as a criticism of a popular (straw) Kuhn.

7　The actual historical picture here is more complex than this suggests. The more sophisticated Positivists tended towards a more

'Wittgensteinian' view. For explanation, see Michael Friedman, *Reconsidering Logical Positivism*.

8 *SSR*, ch. 4; *SSR*, xi; Hung, *The Nature of Science*.

9 For an explication of how so-called Wittgensteinian 'quietism', when understood aright, need not involve actually being quiet about anything, see David Cerbone, 'How to do things with wood', pp. 308–9.

10 Wittgenstein, *Philosophical Investigations*, para. 402.

11 Pickering, *Constructing Quarks*.

12 Wittgenstein, *Philosophical Investigations*, para. 124.

Bibliography

Works by Kuhn

The following are the main works referred to in the text. All page references to *The Structure of Scientific Revolutions* are to the 3rd edition. The earlier editions have the same page numbers for the entire book, excepting only the preface. For a complete Kuhn bibliography, including all his minor and specialist works, please see *The Road since 'Structure'*.

The Copernican Revolution: Planetary Astronomy in the Development of Western Thought (1957), Harvard University Press, 1990.
The Structure of Scientific Revolutions (1962), 3rd edn, University of Chicago Press, 1996.
The Essential Tension: Selected Studies in Scientific Tradition and Change (1977), University of Chicago Press, 1977.
Black-Body Theory and the Quantum Discontinuity, 1894–1912 (1979), University of Chicago Press, 1984.
The Road since 'Structure': Philosophical Essays, 1970–1993, with an Autobiographical Interview, ed. James Conant and John Haugeland (2000), University of Chicago Press, 2000.

Other works cited

Asimov, Isaac, *Understanding Physics*, Dorset Press, 1966.
Barnes, Barry, *T. S. Kuhn and Social Science*, London: Macmillan, 1982.
Barry, Donald, *Forms of Life and Following Rules*, New York: E. J. Brill, 1996.
Baumberger, Jorg, 'No Kuhnian revolutions in economics', *Journal of Economic Issues*, 11:1 (Mar. 1977), 1–20.

Bernstein, Richard, *Beyond Objectivism and Relativism*, Oxford: Blackwell, 1983.

Bird, Alexander, *Thomas Kuhn*, Chesham: Acumen, 2000.

Buchwald, Jed and Smith, George, 'Thomas S. Kuhn, 1922–1996', *Philosophy of Science*, 46:2 (1997), 361–76.

Cavell, Stanley, *The World Viewed*, enlarged edn, Cambridge, Mass.: Harvard University Press, 1979.

Cerbone, David, 'How to do things with wood', in Crary and Read, *The New Wittgenstein*, 2000.

Crary, Alice and Read, Rupert (eds), *The New Wittgenstein*, London: Routledge, 2000.

Currie, Mark, *Postmodern Narrative Theory*, Basingstoke: Macmillan, 1998.

Davidson, Donald, *Inquiries into Truth and Interpretation*, Oxford: Oxford University Press, 1984.

Donovan, Arthur, Laudan, Larry and Laudan, Rachel (eds), *Scrutinizing Science*, Dordrecht: Reidel, 1988.

Duhem, Pierre, *The Aims and Structure of Physical Theory*, Princeton: Princeton University Press, 1954.

Feyerabend, Paul, 'Explanation, reduction and empiricism', in Herbert Feigl and Grover Maxwell (eds), *Scientific Explanation, Space and Time: Minnesota Studies in the Philosophy of Science*, vol. 3, Minneapolis: University of Minnesota Press, 1962.

——*Science in a Free Society*, London: New Left Books, 1978.

——'Some observations on the decay of philosophy of science', in *Philosophical Papers, vol. 2: Problems of Empiricism*, Cambridge: Cambridge University Press, 1981.

——*Against Method*, London: Verso, 1993.

——'Two letters of Paul Feyerabend to Thomas S. Kuhn on a draft of *SSR*', ed. Paul Hoyningen-Huene, *Studies in the History of the Philosophy of Science*, 26:3 (1995), 353–87.

Fleck, Ludwig, *Genesis and Development of a Scientific Fact*, Chicago: University of Chicago Press, 1979.

Friedman, Michael, *Reconsidering Logical Positivism*, Cambridge: Cambridge University Press, 1999.

Fuller, Steve, *Thomas Kuhn: A Philosophical History for our Times*, Chicago: University of Chicago Press, 2000.

Garfinkel, Harold, *Studies in Ethnomethodology*, Englewood Cliffs, N.J.: Prentice-Hall, 1967.

Gill, Jerry, *Wittgenstein and Metaphor*, Atlantic Highlands, N.J.: Humanities Press, 1996.

Glock, Hans-Johann, 'The indispensability of translation', *Philosophical Quarterly*, 43 (1993), 194–203.

Goodman, Nelson, *Ways of Worldmaking*, Hassocks: Harvester, 1978.

Guetti, James, *Wittgenstein and the Grammar of Literary Experience*, Athens, Ga: University of Georgia Press, 1993.

Hacking, Ian, *Why Does Language Matter to Philosophy?* Cambridge: Cambridge University Press, 1975.

——*Representing and Intervening*, Cambridge: Cambridge University Press, 1983.

——'Rules, scepticism, proof, Wittgenstein', in Ian Hacking (ed.), *Exercises in Analysis*, Cambridge: Cambridge University Press, 1985.

——'Working in a new world', in Horwich, *World Changes*, 1993.

——(ed.), *Scientific Revolutions*, Oxford: Oxford University Press, 1981.

Hanson, Norwood Russell, *Patterns of Discovery: An Inquiry into the Conceptual Foundations of Science*, Cambridge: Cambridge University Press, 1958.

Hollis, Martin and Lukes, Steven (eds), *Rationality and Relativism*, Oxford: Blackwell, 1982.

Horwich, Paul, *World Changes*, Cambridge, Mass.: MIT Press, 1993.

Hoyningen-Huene, Paul, 'The interrelations between the philosophy, history and sociology of science in Thomas Kuhn's theory of scientific development', *British Journal for The Philosophy of Science*, 43 (1992), 487–501.

——*Reconstructing Scientific Revolutions*, Chicago: University of Chicago Press, 1993.

Hung, Edwin, *The Nature of Science*, London: Wadsworth, 1977.

Irzik, Gurol and Grunberg, Ted, 'Carnap and Kuhn: arch enemies or close allies?', *British Journal for the Philosophy of Science*, 46 (1995), 285–307.

Koestler, Arthur, *The Sleepwalkers*, London: Hutchinson, 1959.

Kripke, Saul, *Naming and Necessity*, rev. and enlarged edn, Oxford: Blackwell, 1980.

Kuhn, Thomas and Heilbron, John L. 'The genesis of the Bohr atom', *Historical Studies in the Physical Sciences*, 1 (1969), 211–90.

Lakatos, Imre, *Philosophical Papers*, vol. 1, Cambridge: Cambridge University Press, 1978.

Lakatos, Imre and Musgrave, Alan (eds), *Criticism and the Growth of Knowledge*, Cambridge: Cambridge University Press, 1970.

Latour, Bruno and Woolgar, Steve, *Laboratory Life: The Construction of Scientific Facts*, rev. edn, Princeton: Princeton University Press, 1986.

Lee, Keekok, 'Kuhn, a Re-appraisal', *Explorations in Knowledge*, 1 (1984).

McDowell, John, *Mind and World*, Cambridge, Mass.: Harvard University Press, 1994.

——'Towards rehabilitating objectivity', in Robert Brandom (ed.), *Rorty and his Critics*, Cambridge: Cambridge University Press, 2000.

Midgeley, Mary, *The Ethical Animal*, London: Routledge, 1994.

Pickering, Andrew, *Constructing Quarks: A Sociological History of Particle Physics*, Edinburgh: Edinburgh University Press, 1984.

Popper, Karl, *The Myth of the Framework*, London: Routledge, 1994.

——*Realism and the Aims of Science*, London: Hutchinson, 1982.

Preston, John, *Feyerabend*, Cambridge: Polity, 1997.

Putnam, Hilary, *Philosophical Papers, vol. 2: Mind, Language and Reality*, Cambridge: Cambridge University Press, 1975.

—— 'The "corroboration" of theories', in Hacking, *Scientific Revolutions*, 1981.

Read, Rupert, 'On wanting to say "All we need is a paradigm"', *Harvard Review of Philosophy* (2001).

—— 'Understanding Kuhnian incommensurability: some unexpected analogies from Wittgenstein', *Wittgenstein Studies* (forthcoming).

Read, Rupert and Sharrock Wes, 'Thomas Kuhn's misunderstood relation to Putnam/Kripke essentialism', *Journal of the General Philosophy of Science* (forthcoming).

Reza, Yasmina, *Life Times Three*, London: Faber, 2001.

Robinson, Guy, *Philosophy and Mystification*, London: Routledge, 1998.

Rouse, Joseph, *Knowledge and Power*, Ithaca: Cornell University Press, 1987.

Rorty, Richard, 'Being that can be understood is language', *London Review of Books*, 16 Mar. 2000.

Sankey, Howard, *The Incommensurability Thesis*, Aldershot: Avebury, 1994.

Scheffler, Israel, *Science and Subjectivity*, Indianapolis: Bobbs-Merrill, 1967.

Shapere, Dudley, 'The Structure of Scientific Revolutions', *Philosophical Review*, 73 (1964), 383–94.

Slavney, Phillip and Paul, McHugh, *Psychiatric Polarities*, Baltimore: Johns Hopkins University Press, 1987.

Suppe, Frederick (ed.), *The Structure of Scientific Theories*, Urbana: University of Illinois Press, 1977.

Truesdell, Clifford, 'Reactions of late Baroque mechanics to success, conjecture, error and failure in Newton's *Principia*', *Texas Quarterly*, 10 (1967), 238–58.

Williams, Bernard, 'Wittgenstein and Idealism', in *Moral Luck: Philosophical Papers 1973–1980*, Cambridge: Cambridge University Press, 1981.

Winch, Peter, *The Idea of a Social Science*, London: Routledge, 1958 (2nd edn 1990).

—— 'Understanding a primitive society', *American Philosophical Quarterly*, 1 (1964), 307–24.

Wittgenstein, Ludwig, *Tractatus Logico-Philosophicus*, London: Routledge, 1922.

—— *Philosophical Investigations*, Oxford: Blackwell, 1953.

—— *Lectures and Conversations on Aesthetics, Psychology and Religious Belief*, Berkeley: University of California Press, 1970.

—— *On Certainty*, Oxford: Blackwell, 1974.

Index

Lightning Source UK Ltd.
Milton Keynes UK
UKHW010756100622
404202UK00002B/30